AF770943

PROTEOGENOMICS

PROTEOGENOMICS

Mohamed A. Selmy

Delve
PUBLISHING

www.delvepublishing.com

Proteogenomics

Mohamed A. Selmy

Delve Publishing

2010 Winston Park Drive,

2nd Floor

Oakville, ON L6H 5R7

Canada

www.delvepublishing.com

Tel: 001-289-291-7705
001-905-616-2116

Fax: 001-289-291-7601

Email: orders@arclereducation.com

© 2019 Delve Publishing

ISBN: 978-1-77361-159-4 (Hardcover)

This book contains information obtained from highly regarded resources. Reprinted material sources are indicated and copyright remains with the original owners. Copyright for images and other graphics remains with the original owners as indicated. A Wide variety of references are listed. Reasonable efforts have been made to publish reliable data. Authors or Editors or Publishers are not responsible for the accuracy of the information in the published chapters or consequences of their use. The publisher assumes no responsibility for any damage or grievance to the persons or property arising out of the use of any materials, instructions, methods or thoughts in the book. The authors or editors and the publisher have attempted to trace the copyright holders of all material reproduced in this publication and apologize to copyright holders if permission has not been obtained. If any copyright holder has not been acknowledged, please write to us so we may rectify.

Notice: Registered trademark of products or corporate names are used only for explanation and identification without intent of infringement.

Delve Publishing publishes wide variety of books and eBooks. For more information about Delve Publishing and its products, visit our website at www.delvepublishing.com.

ABOUT THE AUTHOR

Mohamed Selmy is a physician scientist and lecturer of Medical Biochemistry and Molecular Biology. He received his medical degree in 2004 from the faculty of medicine Suez Canal University in Egypt. He received his scientific training through many fellowships in United States, Spain and Italy. He received his PhD in Medical Biochemistry and Cancer biology in 2017 from the faculty of medicine Suez Canal University in Egypt. His fields of expertise comprises; cancer biology, cell signaling and molecular medicine. His work is continuously published in international peer reviewed journals.

TABLE OF CONTENTS

LIST OF FIGURES

LIST OF TABLES

LIST OF ABBREVIATIONS

AFLP	Amplification Fragment Length Polymorphism
ANOVA	Analysis of Variance
APEX	Arrayed Primer Extension
AS-PCR	Allele-Specific PCR
ASO	Allele Specific Oligonucleotides
ASPE	Allele-Specific Primer Extension
cDNA	Complementary DNA
CFET	Combinatorial Fluorescence Energy Transfer
CGP	The Cancer Genome Project
CPE	Common Primer Extension
CPs	Cationic Polymers
CSPS	Comparative Shotgun Protein Sequencing
DDBJ	DNA Data Bank of Japan
ddNTP	Dideoxynucleotides
DNA	Deoxyribonucleic Acid
EGFR	Epidermal Growth Factor Receptor
EST	Expressed Sequence Tag
FDR	False Discovery Rate
FP	Fluorescence Polarization
GAPP	Genome Annotating Proteomic Pipeline
HPLC	High Performance Liquid Chromatography-Ultraviolet
HUGO	Human Genome Organization
Ig	Immunoglobulins
KEGG	Kyoto Encyclopedia of Genes and Genomes
LC-MS/MS	Liquid Chromatography Coupled With Tandem Mass Spectrometry
LINE-1	Long Interspersed Nuclear Elements-1
LUMA	Luminometric Methylation Assay
MALDI	Matrix-Assisted Laser Desorption Ionization

MS	Mass Spectrometry
NGS	Next Generation Sequencing
NIH	National Institutes of Health
NIHS	National Institutes of Health Sciences
NMR	Nuclear Magnetic Resonance
ORF	Open Reading Frame
PCR	Polymerase Chain Reaction
PCR-SSCP	PCR- Single-Strand Conformation Polymorphism
PDB	Protein Data Bank
PEP	Posterior Error Probability
PPI	Protein–Protein Interaction
qRT-PCR	Quantitative Real Time Polymerase Chain Reaction
QSAR	Quantitative Structure-Activity Relationship
QTL	Quantitative Trait Locus
RCA	Rolling Circle Amplification
Ref	Reference
RFLP	Restriction Fragment Length Polymorphism
RNA	Ribonucleic Acid
RNA-Seq	RNA Sequencing
SAAVs	Single Amino Acid Variants
SAGE	Serial Analysis of Gene Expression
SAP	Single Amino Acid Polymorphism
SBE	Single Base Extension
SBL	Sequencing by Ligation
SBS	Sequencing by Synthesis
SMART	Single-Molecule Real-Time Sequencing Technology
SNP	Single Nucleotide Polymorphisms
SOPs	Standard Operating Procedures
TCGA	The Cancer Genome Atlas
TIRF	Total Internal Reflection Fluorescence
TOF	Time of Flight
TPP	Trans Proteomics Pipeline
UTR	Untranslated Region

PREFACE

Substantial advancement has been made in our comprehension of the composite relationships between DNA, RNA, and protein, and this progress is being used to both apprehend basic biological process and enhance molecular markers and drug target detection.

It is now quite frequent to observe genomics data informing proteomics data investigation or vice versa. In relation, the fast technological progress taking place over the last decade have made it conceivable to create large proteogenomics data sets, thereby driving the development of new methods for proteogenomics data analysis.

Nevertheless, still many challenges in proteogenomics comprising the incomplete coverage and dynamic range of MS-based proteomics, and the struggle of generating proteogenomics data sets with large numbers of samples as a result of the cost and sample accessibility. Additionally, our knowledge of basic biological routes and pathways is still imperfect, veiling our ability to elucidate many of the proteogenomics relationships we recognize. In spite of all these obstacles, we see a tremendous development in this field both from a data generation side—with better, genomic and proteomic curation and profiling—and the side of data analysis, with the continued development of much better computational tools for data integration and visualization.

This book shades light on basics of proteogenomics. We have started with an illustration of basic knowledge of Omics tools and then we further illustrate the role of bioinformatics in this emerging field. We then provide an illustration of the current utilized software programs in this field. Finally, we dedicated the major part of the application of proteogenomics in selected fields as cancer biology, microbiology and the analysis of antibody repertoires.

CHAPTER 1

INTRODUCTION TO PROTEOGENOMICS

The newly introduced throughput molecular technologies; genomics, proteomics, and metabolomics have augmented the number of detected genes and protein. This, in turn, leads to a better explanation of the expression profiles and networking elucidation.

These "omic" technologies portray and analyze the biological framework through the unraveling of details of a myriad of thousands of genes and proteins. Utilization of these modern technologies remarkably changed the way of understanding of expression networks especially in the medical field and opened several potential areas of research (Ulrich-Merzenich et al., 2007). Such modern technologies comprise the elucidation and clarification, in a non-biased manner, of genomics (which target genes analysis), transcriptomics (targeting mRNA analysis), proteomics (analyzing proteins) and metabolomics (which deals with all metabolites) of a biological sample. Further development of these technologies allows for the integration of more than one technology in a high dimensional biology approach. In relation, system biology, which is a systematic study of various relations and connections in biological systems, represents a model of such integration modalities (Westerhoff and Palsson, 2004; Kell, 2007).

While modern omics technologies are primarily used for a better a clarification of signaling pathways and normal physiological processes, they also play a major role in understanding the etiopathogenesis of many disorders, screening and prognosis of many diseases (Horgan and Kenny, 2011). Pharmacogenomics, which represents a way of interaction between pharmacology and genomics, is a successful application of such technologies in drug discovery and the assessment of drug efficacy and toxicity (Kell, 2006).

To further elucidate, genomics generally represents the throughput large-scale analysis of various genes or a DNA region (Cook-Deegan et al., 2000). In this way, the term genomics comprised the genetics area involving genome sequencing and DNA analysis of an organism. In other words, it is analysis and illustration of the whole DNA content that is present within a single cell of an organism. Further details on genomics and genomic techniques will be discussed in the next chapter.

The accomplishment of the human genome project further energizes the genomics tools. It provided a high throughput tool for gene and DNA analysis and, furthermore, it opened the gates for the throughput analysis of gene expression (Merrill et al., 2006; Bendixen et al., 2010). Based on the knowledge extrapolated and the rise of the microarray technology as a method of synchronized analysis, transcriptomics methods have significantly developed and turned from the analysis of only discrete transcripts into the mass analysis and sequencing of the whole RNAs expressed in living cells. This integration between genomics and transcriptomics, currently called functional genomics, has allowed for a better understanding of the link between the genotype and phenotype and revolutionized the field of modern molecular technology.

It was previously assumed that the presence of a transcript or a phenotypic character is an adequate indicator of gene functionality. Currently, it is well-established that a better indicator of gene functionality is the existence of a functional protein. Therefore, as a result of the need for further integration between functional genomics and protein level investigation, the proteomic technologies have progressively developed. Proteomics represents the high-throughput analysis method to analyze the expression, structure, and function of proteins within a cell or a whole organism (Theodorescu and Mischak, 2007). The classical result of proteomic experiments is the evaluation of the differentially expressed protein content across multiple conditions and among various organs (Silva et al., 2006; Boersema et al., 2015). Proteomic research shows a high feasibility as genome sequences are readily available online for curation and online investigation. The usual experimental plan is the proteolytic breakdown of the protein and the result fragments are assessed by mass spectrometry (MS). The resulting peptide sequences are further evaluated for genomic databases and identification of proteins (Zhang et al., 2013). The use of MS, genomic databases, and sequence homology search are now the cornerstone of all proteomic experimental work. Next in this book, we will deeply review current progress in the field of proteomics.

It should be also highlighted that bioinformatic analysis is the core of integration between the current modern molecular modalities and the groundwork for data assimilation into meaningful results (Schneider and Orchard, 2011). The integrated modern technologies (genomics, transcriptomics, and proteomics) can currently be implemented in all aspects of medical and life science research from screening, accurate diagnosis and personalized therapeutics to the assessment of toxicity, efficient antibiotic production, and mechanisms of resistance. Moreover, it could be used to assess the diversity of organisms, mechanisms of adaptation and response to ecological changes.

One of the further refinements of these modern omic technologies is proteogenomics. The need for proteogenomics arises as a result of the differences between the proteomic approach in comparison to the functional genomic approach. Proteomics differs from functional genomics in two major points. First, the possibility of DNA amplification by technologies such as polymerase chain reaction is not comparably present at the protein level. In this way, prior meticulous purification of proteins appears mandatory for identification and quantification steps. The second major point is that currently there are no such efficient high throughput protein arrays to depend on in further analysis. The difficulty appears as a result of the lack of antibodies against all entire proteins. Even more, the antibody recognition and the affinity change derived by the post-translational modification add to the difficulty to show a solid high throughput protein array. Therefore the inter-relation analysis of proteomic and genomic data was challenging, especially with the presence of 20,000 genes and more than hundred proteins possessing two hundred or more post-translational modifications. Such molecular complexity raises the demand for further improvement of the omic approaches and more refined bioinformatic and computational tools (Schneider and Orchard, 2011).

The proteogenomics approach depends on pooling the data derived from the analysis of the followings: (1) DNA and epigenetic regulations, (2) expression of RNA, and (3) protein and its post-translational modifications and merging these data into a practically suitable and highly develop bioinformatic tools (Nesvizhskii, 2014). The crucial step of proteogenomics is to use the genomic data to synthesize customized database of protein sequences. These sequences, in turn, could be used further to investigate the biological model of interest. The readily available entire genome and RNA transcripts allow for an easier explanation of the sequences of the full count of theoretical proteins present. Meanwhile, shotgun liquid chromatography-tandem mass spectrometry (LC-MS/MS) allows for protein detection against modified protein sequences databases. This inter-relational approach ensures reliable method for identification and quantification of proteins. In addition, it allows for confirmation of gene expression at protein level. While appears as a potentially applicable approach, proteogenomics approach faces many challenges such as the reproducibility, sensitivity and the accuracy of genomics and proteomics studies being conducted. Specifically, the precision and accuracy of both genes and proteins sequencing are the major concern as sequence errors would affect the credibility of clinical and functional study reports. Further refinement of the approach with new computational tools will add more the sensitivity and accuracy of this approach. This book will in the next chapter provide an illustrated review of the proteogenomics approach and the applications based on it.

REFERENCES

1. Bendixen, E., Danielsen, M., Larsen, K., & Bendixen, C. (2010). Advances in porcine genomics and proteomicsda toolbox for developing the pig as a model organism for molecular biomedical research. *Brief Funct Genomics 9*, 208e19.
2. Boersema, P. J., Kahraman, A., & Picotti, P. (2015). Proteomics beyond large-scale protein expression analysis. *Curr Opin Biotech 34*, 162e70.
3. Cook-Deegan, R., Chan, C., & Johnson, A. (2000). World survey of funding for genomics research. Final report to the global forum for health research and the world health organization. September, Washington D.C
4. Horgan, R. P., & Kenny, L. C. (2011). 'Omic' technologies: genomics, transcriptomics, proteomics and metabolomics. *Obstet Gynaecol 13*, 189e95.
5. Kell, D. B. (2006). Systems biology, metabolic modelling and metabolomics in drug discovery and development. *Drug Discov Today 11*, 1085e92.
6. Kell, D. B. (2007). The virtual human: towards a global systems biology of multiscale, distributed biochemical network models. *Iubmb Life 59*, 689e95.
7. Merrill, S. A., Mazza, A. M., & in Genomic, P. R. (2006). Genomics, Proteomics, and the changing research environment. Washington (DC): National Academies Press (US)
8. Nesvizhskii, A. I. (2014). Proteogenomics: concepts, applications and computational strategies. *Nat Methods 11*, 1114e25.
9. Schneider, M. V., & Orchard, S. (2011). Omics technologies, data and bioinformatics principles. *Methods Mol Biol 719*, 3e30.
10. Silva, J. C., Denny, R., Dorschel, C., Gorenstein, M. V., Li, G. Z., Richardson, K., Wall, D., & Geromanos, S. J. (2006). Simultaneous qualitative and quantitative analysis of the Escherichia coli proteome, a sweet tale. *Mol Cell Proteomics 5*, 589e607.
11. Theodorescu, D., & Mischak, H. (2007). Mass spectrometry based proteomics in urine biomarker discovery. *World J Urol 25*, 435e43.
12. Ulrich-Merzenich, G., Zeitler, H., Jobst, D., Panek, D., Vetter, H., & Wagner, H. (2007). Application of the "-Omic-" technologies in phytomedicine. *Phytomedicine 14,* 70e82.

13. Westerhoff, H. V., & Palsson, B. O. (2004). The evolution of molecular biology into systems biology. *Nat Biotechnol 22*, 1249e52.
14. Zhang, Y., Fonslow, B. R., Shan, B., Baek, M. C., & Yates, J. R. (2013). Protein analysis by shotgun/bottom-up proteomics. *Chem Rev 113*, 2343e94.

CHAPTER 2

MODERN OMIC TECHNOLOGIES

CONTENTS

2.1. GENOMIC TECHNOLOGIES

As of the last three decades, a shift in the outlook toward genome sequencing was introduced to the molecular field. Data extrapolated from genome sequencing can give a wide comprehension of gene locations, chromosomal information and many other important uses up to offering reference gene maps. The following topics in this chapter illustrate the current technologies used in the field of genomics.

2.1.1. SNP Analysis

Single Nucleotide polymorphisms (SNPs) appear as the most widely occurring genetic variations among organisms (Consortium, 2005, Rocha, Gut et al., 2006). A variation in a single nucleotide between two individuals is only considered as a polymorphism when the base variation is found to be occurring in more than 1% of the total population as this indicates that this base difference is stable in the population (Brookes, 1999). It has been established that SNPs occur in a frequency of one in 1000 bp in the human genome (Wang, Fan et al., 1998; Sachidanandam, Weissman et al., 2001) and these SNPs usually unevenly distributed throughout the genome. These SNPs are regularly biallelic but occasionally they can be also multi-allelic (Kim and Misra, 2007) and most of them are usually found on the non-coding regions of the DNA thus do not modify the encoding proteins (Li and Sadler, 1991; Nickerson, Taylor et al., 1998). Such SNPs are termed as non-synonymous SNPs and could be used as genetic markers for evolution tracking. On the other hand, SNPs in the coding regions result in changes in the coded proteins. Similarly, SNPs occurring in the regulatory regions of the DNA modify the transcription and translation rates.

Most of the technologies for SNP analysis necessitate PCR amplification where a SNP region of interest is amplified with PCR to increase specificity as well as to increase the molecule count. SNP analysis typically comprises two steps; the generation of allele-specific products for Allelic discrimination and then the step of allele determination.

2.1.1.1. Allele Discrimination Techniques

In order to discriminate the alleles, specific biochemical reactions are implemented using methods such as primer extension (nucleotide incorporation), hybridization, ligation, and enzymatic cleavage. Each of these methods is illustrated below in detail.

A) **Allelic discrimination by Primer extension:** This method allows for an allelic discrimination through performing a primer extension reaction on a DNA template where specific nucleotides anneal to the complementary strand. In this way, either a solitary common primer or allele specific primers are used to discriminate the alleles. Most of the approaches make use of single primer for both the alleles. A common primer extension reaction can detect many SNPs simultaneously and is conducted in the following steps (Sokolov, 1990); annealing of the designed primer with the 3' end near to the SNP of interest. Then, an extension step is done using ddNTPs and polymerase enzyme followed by the identification of the extended base.

Assays using mass spectrometry: In a common primer extension (CPE) reaction (Figure 2.1a) primer and ddNTPs are used where the primer anneals one base upstream to the SNP site of interest. After the annealing, extension occurs with the help of available ddNTPs. Thus extended products are then detected for genotypes based on the differences between mass of extended product using mass spectrometry. Assays such as PinPoint assay (Haff and Smirnov, 1997, Ross, Hall et al., 1998), GOOD assay (Sauer, Gelfand et al., 2002), MassEXTEND[tm] (Braun, Little et al., 1997, Cashman, Zhang et al., 2001), and SPC-SBE (Kim, Edwards et al., 2002) assays use the above CPE-based methods for allelic discrimination. MALDI-TOF MS (matrix assisted laser desorption/ionization time-of-flight mass spectrometry) is usually used for the detection of these discriminated alleles (Figure 2.1a). Here PCR products are used as templates where allele specific primers aid in the extension using ddNTPs resulting in products with different masses. These extended products represent one of the alleles and are further evaluated for their genotype. Assays that use mass spectrometry are illustrated in Table 2.1.

Table 2.1. Assays Implementing Mass Spectrometry

Assays	**Protocol of the assay**
PinPoint assay	It is considered to be the simplest of all performs a single base extension (SBE) after the annealing of the primer. No modifications of the nucleotides or the primers are done.
MassEXTEND[tm]	Employs high throughput genotyping on MassARRAYtm (Sequenom,CA) platform in an automated manner.

GOOD assay	Primers are altered with phosphorothioate at their 3 ends. Single base extension (SBE) is performed using α-S-ddNTPs yielding charge-tagged DNA fragments. The extended products need not be purified for further MS analysis because of the charge tagging. In this way, the spectral resolution achieved is much higher because of the use of short extended fragments.
SPC-SBE assay	In this technique, ddNTPs attached to a biotin moiety (biotin-ddNTPs) are used. The extended products are 3-biotinylated DNA fragments which can be separated from the reaction mixture using streptavidin-coated beads. This method of isolation using molecular affinity helps in multiplexing where unextended primers that overlap with extension products are discarded in the mass spectrum. Elimination of salts also results in the enhancement of the mass spectral peak resolution.

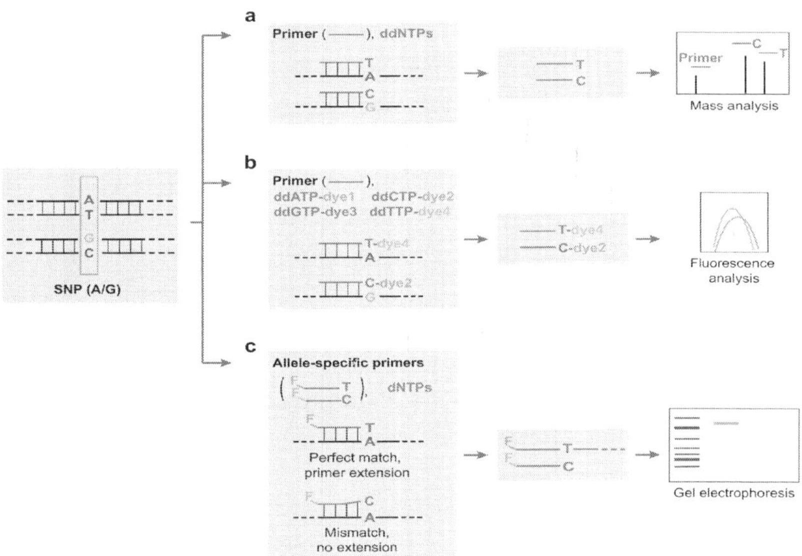

Figure 2.1. Primer extension approaches for SNP genotyping. (a) Mass spectrometry–based detection. (b) Fluorescence detection using capillary electrophoresis. (c) Allele-specific primers with detection of PCR products (Kim and Misra, 2007).

Assays using fluorescence based detection: These approaches execute fluorescence detection employing capillary electrophoresis (Figure 2.1b), where fluorescent tagged ddNTPs are used in place of normal ddNTPs. The detection of SNP genotype is based on the color of the dye attached to the specific ddNTP. Various CPE approaches that use fluorescence-based detection are illustrated in Table 2.2.

Table 2.2. Description of Primer Extension Assays That Implement Fluorescence Based Detection

Assays using fluorescence based detection	**Protocol of the assay**
SNaPshot® approach (Applied Biosystems, CA)	This method employs a homogeneous reaction with PCR amplicons, primers, and fluorescently labeled ddNTPs. In this approach, a single base extension is performed using these fluorescently labeled ddNTPs which are then detected by capillary array electrophoresis (Le Hellard, Ballereau et al., 2002). About 35 SNPs were discriminated and detected using this technique earlier. This technique allows for genotyping of about 95 samples simultaneously (Sanchez, Børsting et al., 2003).
SNP stream^{tm} assay (Orchid Biosciences, NJ)	This assay employs a different approach where tagged primers are used in a homogenous reaction. The final extended products are captured onto a solid support with the help of an array comprising complementary tags (Fan, Chen et al., 2000, Hirschhorn, Sklar et al., 2000). After capturing the products, array is then washed and then scanned to detect the fluorescence signal. Genotype is determined based on the fluorescence signal emitted by the tags on the array. In this unique assay, single base extension (SBE) is performed using a 384 well plate where each well has 12-tagged primers (Nikiforov, Rendle et al., 1994, Bell, Chaturvedi et al., 2002). Arrayed primer extension (APEX) assays use solid phase reactions have primers with their 5' ends attached to a solid support (Pastinen, Kurg et al., 1997). After hybridization of the templates to the primers, genotype determination is performed. Another hybrid approach is to make use of both PCR as well as SBE reactions using same primers on a solid support (Shapero, Leuther et al., 2001). Using a solid support is more advantageous as this allows for reducing the interactions between the primers thus increasing the specificity.

Specific primer extension (SPE) approaches: These approaches use instead of a single primer, an allele specific primers (Figure 2.1c) that anneal with the 3' end at the SNP site. These primers only differ at their 3' end and are identical otherwise. Accordingly, extension of a base happens only when the end is complementing the SNP on the template resulting in the discrimination of the allele. The following assays in Table 2.3 are employing this approach.

Table 2.3. Description of Primer Extension Assays that use Specific Primer Extension (SPE) Approaches

Allele-specific PCR (AS-PCR):	In this method, PCR amplification of genomic DNA is executed using differently labeled allele-specific primers and a common reverse primer. Allele-specific primer only binds to the SNP site which is complementary to it, therefore a product is obtained only in these cases (Gibbs, Nguyen et al., 1989). This approach can also be implemented in combination with real-time PCR and capillary array electrophoresis (Medintz, Wong et al., 2001, McClay, Sugden et al., 2002, Latorra, Campbell et al., 2003).
Allele-specific primer extension (ASPE)	In this method, along with the PCR amplified template, allele-specific primer extensions are used. Accordingly, extended products are then used for genotype determination by the detection of fluorescence (Ugozzoli, Wahlqvist et al., 1992).
Tag-Ittm approach (Tm Bioscience Corp., Canada)	Here, tagged primers along with labeled nucleotides are used for ASPE. The extended products are tagged which are then captured via the anti-tag carrying fluorescently labeled microbeads. Further analysis using flow cytometry helps in the identification of the SNP genotype (Bortolin, Black et al., 2004).

B) **Allelic discrimination by Hybridization;** Hybridization methods allow for allelic discrimination by the distinction between the perfectly matched from the mismatched target-probe pairs based on the variances in the thermal stability of double stranded DNA. SNP location in the probe, Probe sequence and length, conditions for hybridization etc., impact the effectiveness of the approach. As this approach does not comprise any enzymatic reactions, it is the favored approach for allele discrimination by high throughput platforms using microarrays (Table 2.4).

Table 2.4. Description of Primer Extension Assays That Use Hybridization-Based Approaches

Hybridization in combination with	Technology	Description
Fluorescent signal detection	GeneChip® array technology (Affymetrix, CA)	Photolithography is used to synthesize allele-specific probes of 25-mer oligonucleotides onto a solid surface. This forms a probe array which is then hybridized with the amplified and tagged SNP-containing genomic region. After hybridization, arrays are washed and the probe-target pairs are analyzed for fluorescent labeling to identify the SNP genotype (Matsuzaki, Loi et al., 2004). This approach allows for parallel genotyping of SNPs with the help of arrays holding millions of probes (Matsuzaki, Dong et al., 2004).
	TaqMan® genotyping assay (Applied Biosystems, CA)	This approach uses 5' polymerase activity in addition to the hybridization and fluorescence detection as depicted in the Figure 2.2b. This assay enables the detection of both the SNPs using a single enzymatic step. In this approach, two allele-specific probes which differ in single base with one end tagged with a fluorescent dye (reporter end) and the other end tagged with a nonfluorescent dye (quencher end) in addition to two primer oligonucleotides that flank the SNP containing region are used. A polymerase with 5' endonuclease activity cleaves the reporter end of the perfectly matched probe-target pairs during the primer extension step of PCR (Holland, Abramson et al., 1991). The emitted fluorescent signal by the reporter dye in the absence of its quencher is then detected to determine the SNP genotype.
MALDI-TOF MS detection	Qbead™ approach (Quantum Dot, CA)	Here, allele specific hybridization is performed through the use of peptide nucleic acid probes (PNA) instead of the traditional probes, as they are more stable and resistant to fragmentation (Griffin, Tang et al., 1997). This difference makes them more suitable during the MALDI-TOF MS analysis. Biotinylated primers are employed to amplify the SNP containing regions followed by hybridization of these PCR products with allele-specific probes found on the quantum dot-encoded microspheres (Xu, Sha et al., 2003). The Hybridized amplicons emit the flourescene signal after the fluorophore-attached streptavidin conjugates with the biotin on the primers and this is detected with the help of a flow cytometer. In this way, a fluorophore signal indicates a successful hybridization and the identity of the SNP is revealed by the unique quantum dot encoding the microsphere.
	Dynamic allele-specific hybridization (DASH™; DynaMetrix, UK)	In this approach, PCR products are separated out based on the streptavidin-coated beads using the biotinylated primers used during PCR amplification. This step is followed by the hybridization of these amplicons with the complementary allele-specific probes in the presence of an intercalating fluorescent dye (Howell, Jobs et al., 1999). The Real-time denaturation of the probe-target pair is then observed by raising the temperature gradually which then gives out a fluorescence-based melting curve. Allelic variants yield different patterns of this curve (Russom, Haasl et al., 2006).

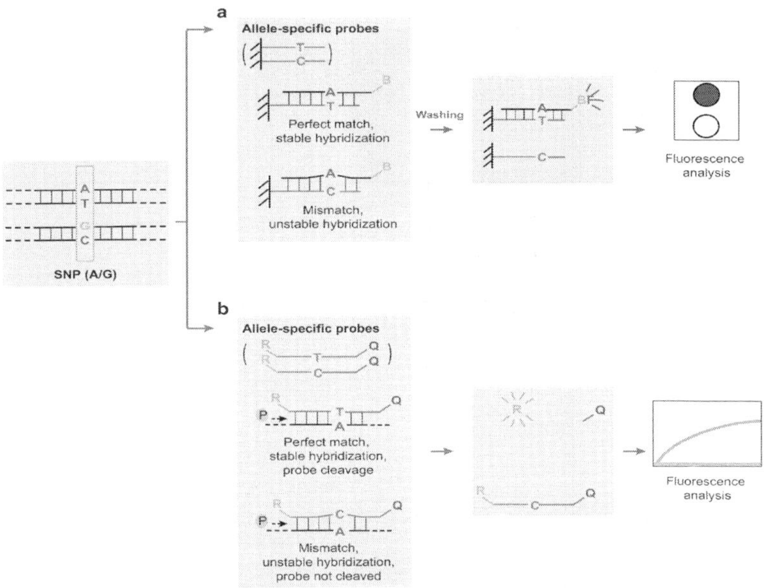

Figure 2.2. Hybridization-based approaches for SNP genotyping. (a) Target hybridization to a probe array. (b) TaqMan® assay (Kim and Misra, 2007).

c) **Allelic discrimination by ligation*:*** In this approach, allelic discrimination is attained through the use of specific ligase enzymes for ligating two oligonucleotides that perfectly bind to the template. In case of three oligonucleotide probes, one is a common probe while the other two are allele-specific (Landegren, Kaiser et al., 1988). Accordingly, ligated products are then analyzed for defining the genotype using allele detection techniques.

Detection using CFET labels: In this approach, two differently labeled allele-specific probes with CFET (combinatorial fluorescence energy transfer) tags are implemented in combination with a biotin-tagged common probe in this assay (Figure 2.3a). In this way, when the allele-specific probes hybridize to their complementary SNP site target template, a specific ligase then ligates these probes to the common probe (Tong, Li et al., 2001). Thus, ligation to the common probe does not happen when there is a mismatch. Biotin tag on the common probe helps in separating these ligated products which are then analyzed using capillary array electrophoresis for defining the genotypes based on the emitted fluorescence signals by the dyes on the CFET tags.

Padlock technology: In this method, a circular strand is formed after ligation of the allele specific probes with the common probe as shown in the Figure 2.3b. In this way, ligation occurs only when there is a perfect match between the probe and the target SNP site (Nilsson, Malmgren et al., 1994). Rolling circle amplification (RCA) or conventional PCR amplification with fluorescently labeled primers are then used to detect the padlock probe (Lizardi, Huang et al., 1998). A single base extension at the SNP site followed by ligation into a modified padlock probe is seen in case of MIP (ParAllele Biosience, CA), the molecular inversion probe technique (Hardenbol, Banér et al., 2003). Unbound probes are then degraded and the bound ones are linearized and amplified with PCR. PCR products are then used for determining the base by hybridizing onto a microarray followed by fluorescence detection. This technology can detect up to 10^4 SNPs accurately (Hardenbol, Yu et al., 2005).

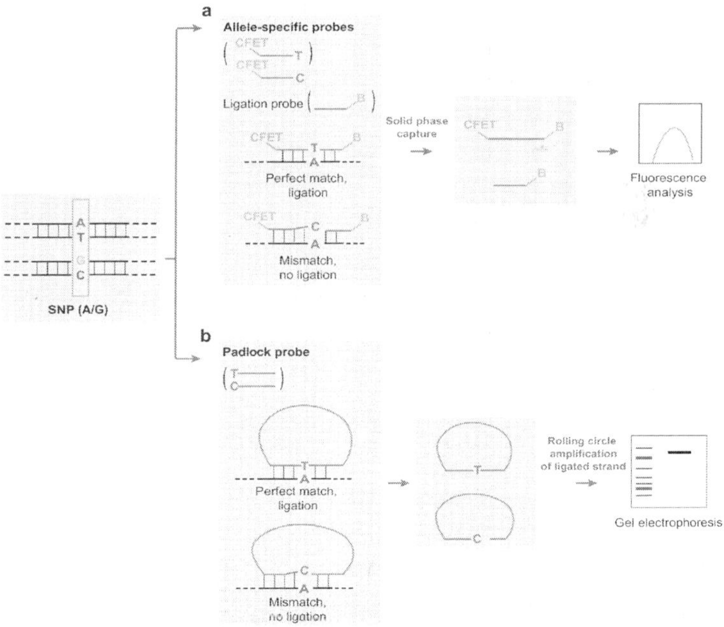

Figure 2.3. Ligation-based SNP genotyping approaches using ligase enzymes. (a) Detection using CFET labels. (b) Ligation of Padlock probe (Kim and Misra, 2007).

Allelic discrimination by enzymatic cleavage: Here DNA cleaving enzymes are used to discriminate alleles present in the enzyme recognition sequence.

RFLP (restriction fragment length polymorphism): This method uses a PCR reaction where enzymes that recognize specific sequences with a SNP on the double stranded DNA and cleaves both the strands either at a specific site or a site nearby to it are used. This result in shorter fragments, which are then determined based on their size using a traditional gel (Botstein, White et al., 1980). Despite it is a simple approach short of any use of probes, it is limited in terms of number of SNPs and throughput analysis.

The Invader® assay (Third Wave™ Technologies, WI): Two allele specific probes that complement to the 5' region of the target template with an overhang and a common probe called invader that complements the 3' region are used for hybridization (Lyamichev, Mast et al., 1999). In case of a perfect match, a three-dimensional structure is formed which is then recognized by cleavase, a flap endonuclease. This enzyme then cleaves the overhang and the resulting fluorescence signal is detected to determine the base, while in case of a mismatch, no overhang is cleaved. This technique allows for the detection of both the allele variants. A modified version of this assay employs the released overhang oligonucleotide as an invader in a second reaction to offer greater signal amplification. This also skips the PCR amplification step and directly detects the SNPs based on the zeptomolar quantities of genomic DNA (Hall, Eis et al., 2000).

2.1.1.1.1. Other Methods for Allelic Discrimination

PCR-SSCP (single-strand conformation polymorphism): This technique is based on the concept that strands with a single base change differ in their native conformations and therefore have different mobility when running on a gel (Orita, Suzuki et al., 1989). A radioisotope labeling or dye staining in combination with a polyacrylamide gel electrophoresis was initially used to detect the mobility of the strands. Capillary array electrophoresis has now replaced the traditional polyacrylamide gel electrophoresis, where strands are labeled fluorescently and then detected on a capillary gel. This method can allow for up to four color labeling and genotyping of multiple SNPs simultaneously in addition to detection of unknown SNPs (Kukita, Higasa et al., 2002).

MRD (Affymetrix, CA): This method is based on the bacterial mismatch repair system, this method detects the SNPs by cloning the PCR amplicons into a vector with an active Cre gene and then growing these vectors in a modified E. coli strain (Faham, Baharloo et al., 2001). Amplification of the identical genomic regions is done with PCR and then amplicons are incorporated into heteroduplexes using vector carrying an inactive Cre gene and unmethylated single-stranded standards. Degradation of the unmethylated strand which possess the mismatch takes place thus losing the active Cre gene. Accordingly, only the strands with proper match have active Cre gene. This technique is used to isolate cells using selective media and then analyzed for detecting the variant using restriction digestion and gel electrophoresis. (Figure 2.4)

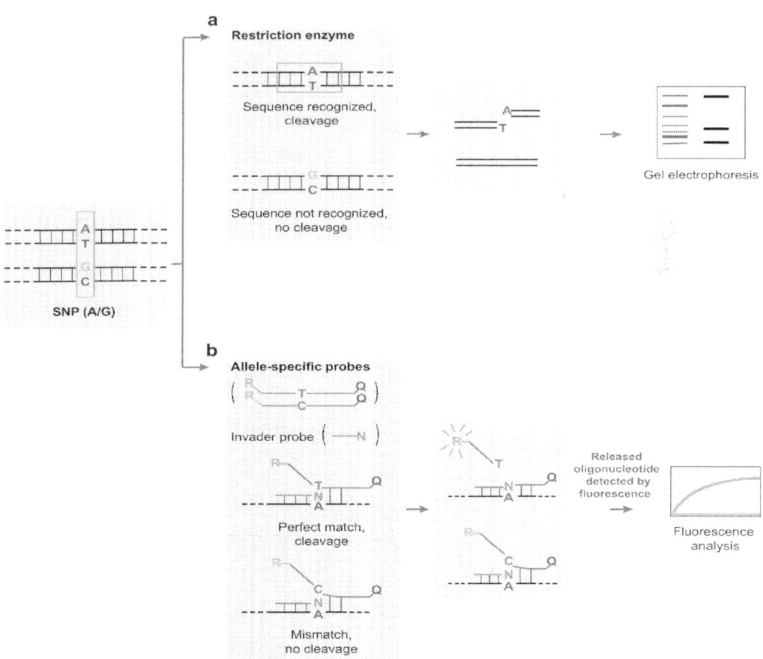

Figure 2.4. Enzymatic cleavage for SNP genotyping. (a) Cleavage with a restriction enzyme. (b) Invader® assay (Kim and Misra, 2007).

Bead Array™ (Illumina, CA): Through the use of this platform, high-throughput SNP genotyping can be conducted for allele discrimination by using a combination of hybridization, primer extension, and ligation followed by PCR amplification. This technique employs a microbead-based array with an optical fiber bundle with around 50,000 individual fibers (Ferguson, Steemers et al., 2000). Each microbead attached covalently to thousands of copies of a particular oligonucleotide sequence is used for the detection of a complementary SNP. Two allele specific oligonucleotides (ASO) and one locus specific oligonucleotide (LSO) are used along with the template of SNP site attached to a solid support for hybridization. Extension of ASO to LSO and ligation are followed after the unbound primers are washed away. PCR amplification of the ligated strand using ASO and LSO specific primer is performed. This is followed by the hybridization of these PCR products with their complementary oligonucleotides attached on a microarray Finally genotype is determined based on the fluorescence signal emitted by the tags carried by the ASO. In this way, a single array can detect up to 150,000 genotypes (Shen, Fan et al., 2005).

SNP lex™ (Applied Biosystems, CA): This approach is quite similar to Bead Array™ (Tobler, Short et al., 2005) in which LSO specific primers here carry a biotin tag. The product strands after ligation are separated using the biotin tag. Hybridization of these products with allele-specific and fluorescent tagged probes is then conducted followed by the genotype identification using capillary array electrophoresis. This method allows for a simultaneous analysis of about 45 SNPs from 96 samples.

2.1.1.2 Allele Detection Methods

After Allele discrimination, the following step is to detect the genotype of the allele. Technologies such as mass spectrometry, fluorescence, and chemiluminescence are used for the detection purpose.

Mass-based detection: With the help of MALDI-TOF MS mass analysis of biomolecules can be achieved, where a matrix (a small organic molecule that ionizes by absorbing energy at a certain wavelength from a laser source) is co-crystallized with a mixture of analytes for ionization (Karas and Hillenkamp, 1988).

Ions generated from this step are then separated based on the time of flight. Time of flight of an ion depends on its mass and charge. This method is implemented by many assays to detect the genotype accurately (Figure 2.5).

Fluorescence signal-based detection: This method of allele detection is highly sensitive as well as fast thus it is most widely used methods for determining the SNP base. Used in combination with capillary gel electrophoresis where the primer extension products are separated out by electrophoresis and the identity of the SNP is then determined based on the specific signal emitted from the product.

Fluorescence polarization (FP)-based detection: This method detects the change in the polarization of a light emitted by fluorescent dye molecule. This change can be due to change in its molecular weight under conditions of constant temperature and solvent viscosity [53]. TaqMan® and Invader® use this method for allele determination.

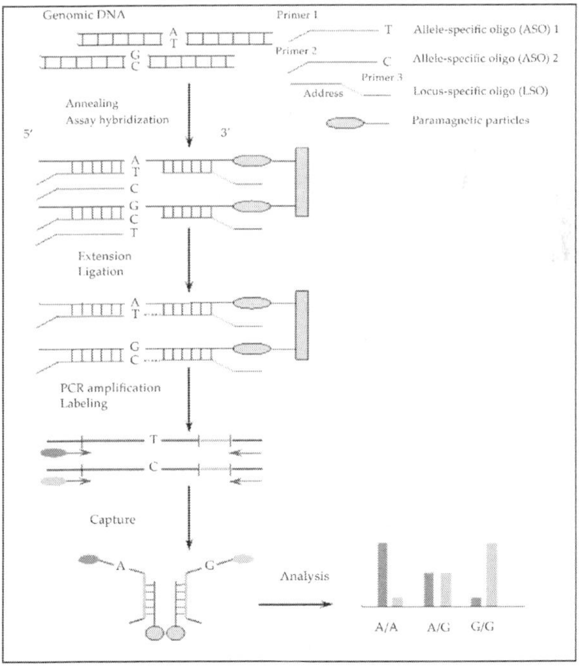

Figure 2.5. Methodology of bead array™ (Illumina, CA) (https://www.ncbi.nlm.nih.gov/probe/docs/techbeadarray/).

Chemiluminescence: This method is employed by Pyrosequencingtm (Biotage, Sweden) to detect the genotype of SNP owing to its several advantages such as rapid detection, high signal-to-noise ratio and feasibility for automation (Ronaghi, Karamohamed et al., 1996). During the genotyping process, after the primer anneals to the template, individual dNTPs are added one by one to the reaction mixture. When there is exact match with the template strand, a light signal is emitted which is then monitored to reveal the genotype. This approach has found to have limited use during multiplexing. (Figure 2.6)

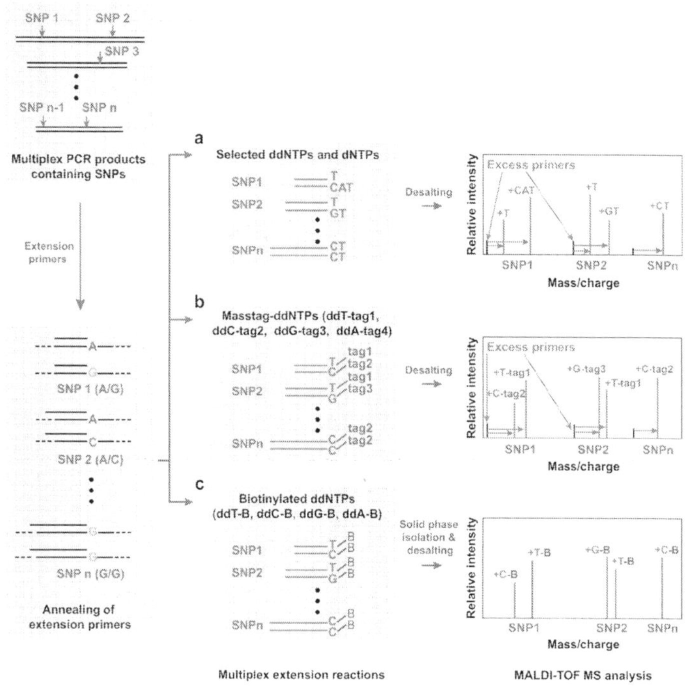

Figure 2.6. Multiplex SNP genotyping by primer extension and MALDI-TOF MS (Kim and Misra, 2007).

2.1.2. Epigenetics Analysis

Apart from the analysis of the genome sequence information, it is also important to investigate the DNA modifications to comprehend the regulation of gene expression in an organism. Epigenetics studies allow for a better understanding of the occurrence of reversible as well as heritable changes in gene function. These alterations are usually due to external sources and take place without modifying the genome sequence itself. These are considered to be major key players in the regulation of gene expression. These changes or modifications are referred to as epigenetic modifications and they include DNA methylation, histone modifications, imprinting, etc.

2.1.2.1. Techniques to Identify DNA Methylation

DNA methylation is the most studied epigenetic modification since it is easily measurable and efficient. DNA methylation is the addition of methyl groups to 5-carbon of cytosine by a DNA methyltransferase. Methylation usually occurs in special regions of DNA called CpG islands which are rich hubs of cytosines and guanines. CpG islands are usually sequences of small stretches of about 300–3000 bp with >50% GC content. While ~70% of promoter regions contain CpG islands, only 1% of the remaining genome has them. DNA methylation can be analyzed at the global or gene specific levels. In case of global analysis, techniques such as chromatography or methyl accepting capacity assays were initially used to measure the levels of methyl cytosines in the whole genome. While in case of gene-specific methylation, Southern blot analysis or PCR amplification were used earlier to identify the regions that were resistant to the methylation-sensitive restriction enzymes.

Currently there many techniques available for studying DNA methylation. Among these techniques it is important to choose the right technique and this can be decided based several factors which include biological question that needs to be answered (Kurdyukov and Bullock, 2016) for example, if a new epigenetic change has to be discovered or already known methylation sites in genes of interest have to be investigated, etc.; quality and amount of DNA sample that is available with the researchers for analysis; How much sensitive or specific the study needs to be? And how simple and robust the employing method has to be?

> *Whole genome methylation:* It usually occurs either when the DNA (de) methylation machinery is altered by any events or cells (Arakawa, Watanabe et al., 2012, Kang, Piao et al., 2014) are treated with compounds such as furan or azacytidine (Conti, Kobets et al., 2014) or learning induced changes in the brain tissue (Gaglio, Capitano et al., 2014) or changes leading to tumor generation (Bujko, Musialik et al., 2014, Calvo, Nomdedeu et al., 2014). The following are currently available and widely used

techniques for studying methylation at the whole genome level (Figure 2.7).

HPLC-UV (high performance liquid chromatography-ultraviolet): This technique is the most preferred and the considered as a gold standard when compared to other methods. Kuo and colleagues have developed this method in 1980 (Kuo, McCune et al., 1980). This method requires large amounts of hydrolyzed DNA sample in addition to specialized laboratory equipment which are considered to be major drawbacks. This method requires hydrolysis of the DNA sample followed by the separation of methylated cytosines (5 mC) and deoxycytidine (dC) using chromatography. The 5 mC/dC ratio is calculated after measuring the fractions of the separated nucleotides. This procedure is performed on both control as well as experimental samples and then the resulting ratios are compared with each other.

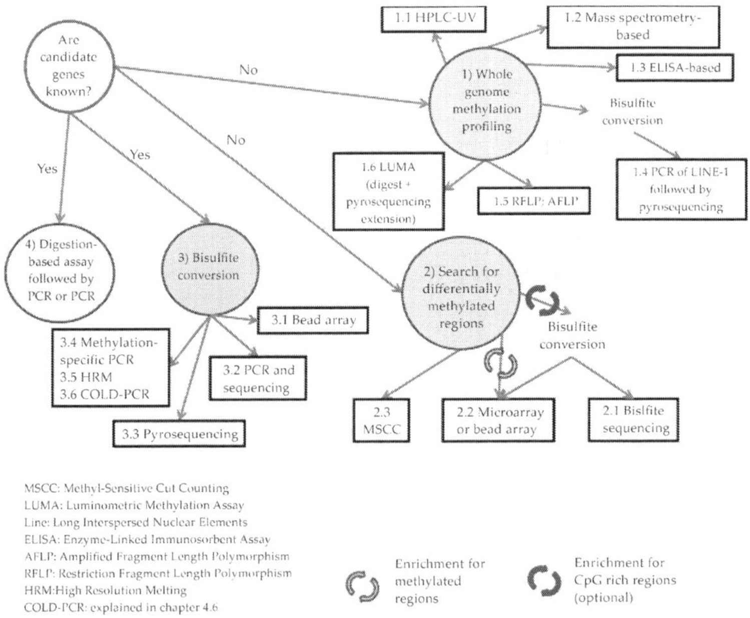

Figure 2.7. Whole Genome Methylation analysis methods (Kurdyukov and Bullock, 2016).

LC-MS/MS Liquid chromatography coupled with tandem mass spectrometry (LC-MS/MS): This method is more accurate and demands lesser samples of hydrolysed DNA when compared to HPLC-UV. But this needs access to LC-MS/MS (Song, James et al., 2005).

ELISA-Based Methods

These methods give only approximate amounts of DNA methylation but are easy and quick; and so are suitable for identifying methylation at the whole genome level. Currently, many ELISA kits are available to perform a quick whole genome level methylation analysis (Table 2.5).

Table 2.5. ELISA Kits for a Quick Whole Genome Level Methylation Analysis

Company	Kit name	Amount of DNA Required
Cell Biolabs	Global DNA Methylation ELISA	2 μg
Sigma-Aldrich	Imprint Methylated DNA Quantification kit (sandwich ELISA)	100 ng
abcam	EpiSeeker methylated DNA Quantification Kit	100 ng
Active Motif	Global DNA Methylation Assay — LINE-1	1 μg
Zymo Research	5-mC DNA ELISA Kit	100 ng
Epigentek	MethylFlash Methylated DNA5-mC Quantification Kit (Colorimetric)	100 ng
	MethylFlash Methylated DNA5-mC Quantification Kit (Fluorometric)	100 ng

In this method, ELISA plates are used to capture the DNA sample and then incubation of these plates with a primary antibody against 5 Mc is performed. Another round of incubation is performed using a secondary antibody followed by the detection using colorimetry or fluorometry.

LINE-1 + Pyrosequencing

Methylation levels in LINE-1 (long interspersed nuclear elements-1) retrotransposons can be analyzed with a combination of bisulfite conversion of DNA and amplification LINE-1 conserved sequences using PCR. These amplicons are then checked for their methylation status using pyrosequencing. This method is known to be efficient with very less DNA samples as well and is used for high throughput analysis (Ogino, Nosho et al., 2008).

AFLP and RFLP

Classical methods such as amplification fragment length polymorphism (AFLP) (Aung, Harrison et al., 2010) and restriction fragment length polymorphism (RFLP) (Jaligot, Beulé et al., 2002) can be used for identifying the methylation levels. With the availability of more advanced techniques these methods are not much preferred these days. Advantages of these methods are that they are very economical and can be used for any species with limited sequence information.

LUMA (luminometric methylation assay)

Karimi and colleagues have come up with this method where two restriction enzymes CpG methylation-sensitive enzyme HpaII and methylation-insensitive enzyme MspI are used to simultaneously in two different reactions to digest all CCGG sites generating 5′-CG overhangs (Karimi, Johansson et al., 2006). Additionally, enzyme EcoRI generating 5′-AATT overhangs is used as a control in both the reactions. Pyrosequencing-based extension assay is then used to fill these overhangs. The ratio of HpaII/MspI is calculated by measuring the light signals and this ratio gives an estimate of the unmethylated DNA in the sample. Advantages of this method include high specificity, low variability, and usage of very fewer amounts of DNA sample.

2.1.2.2. Identification of Differentially-Methylated Regions

The following techniques are used for identifying and assessing specific genes/regulatory regions of interest that are differentially methylated.

Bisulfite Sequencing: This is one of the most widely used methods for determining the differentially methylated genes. Treating the DNA with bisulfite will result in the deamination of all cytosines, except 5-methylcytosine, into uracil. This way methylated cytosines can be distinguished from the normal cytosines. During sequencing of bisulfite treated samples, uracil will be read as thymine instead, therefore, making it possible to distinguish normal cytosines from the methylated ones. This can also be extended to the whole genome level with the help of NGS technologies. Whole genome bisulfite sequencing (WGBS) has an additional step of bisulfite conversion prior to the usual whole genome sequencing protocol. Major drawbacks of this method include complex downstream analysis of sequencing as only three nucleotides are now read instead of four. Also DNA fragmentation because of bisulfite can end up generating shorter fragments which ultimately might result in chimeric products.

Sequencing only a part of the genome which is enriched with the methylating regions, i.e., reduced representation bisulfite sequencing (RRBS) (Meissner, Gnirke et al., 2005, Lee, Jin et al., 2014) can be more affordable and give accurate results as well. Many platforms such as Illumina, Life Technologies etc. offer these services. For this, CpG rich regions have to be separated first which is done by digesting the DNA with MspI digestion that recognizes CCGG sites. This enzyme cuts both methylated as well as non-methylated regions. About ~85% of CpG islands in the human genome can be isolated using this technique. After the separation of these regions, bisulfite conversion followed by sequencing is performed. Another technique is to enrich these CpG regions before bisulfite conversion. This can be done by hybridizing these regions with bait sequences which are immobilized oligonucleotides. Commercial kits such as SureSelect Human Methyl-Seq from Agilent could be used for this purpose. Few kits such as SeqCap Epi CpGiant Enrichment Kit from Roche also provide hybridization after bisulfite conversion. Targeted bisulfite sequencing allows enrichment of only small regions of interest (Walker, Bhagwate et al., 2015).

Pacific Biosciences, on the other hand, has a protocol for detecting the methylated bases directly through sequencing without any enrichment or bisulfite conversion. This protocol involves single cell sequencing where the polymerase kinetics are monitored for detecting the methylated bases (Flusberg, Webster et al., 2010). Another proposed technology namely nanopore-based single-molecule real-time sequencing technology (SMRT) is also working on detecting these methylated bases directly (Laszlo, Derrington et al., 2013) (Schreiber, Wescoe et al., 2013).

Array or Bead Hybridization: Another technique for detecting the 5-methylcytosine is by using microarrays after separating the methylated fragments of DNA by immunoprecipitation. Few commercially available kits such as Human CpG Island Microarray Kit (Agilent), the GeneChip Human Promoter 1.0R Array and the GeneChip Human Tiling 2.0R Array Set (Affymetrix) can be used for this purpose. Few kits such as Infinium Human Methylation450 Bead Chip array by Illumina focus only on certain regions such as gene promoter regions, enhancer regulatory elements and 31 untranslated regions (31UTRs) where methylation usually occurs. This can be cost-effective as well as time saving. This particular array has the ability to detect the methylation status of about 485,000 individual CpG from known genes such as miRNA promoters, 51 UTR, 31 UTR, coding regions (~17 CpG per gene) and island shores (regions ~2 kb upstream of the CpG islands). This need DNA sample of about ~500 ng. The protocol of this kit includes the following steps:

Genomic DNA treated with bisulfite.

Then oligos complementary to uracil and cytosine are added to the reaction mixture.

Hybridization followed by extension and ligation of primers to locus specific oligos.

Creation of detectable products using labeled PCR primers.

Immobilization of these products onto bar-coded beads.

Measurement of signal and calculation of methylation level based on the ratio between two types of beads.

Methyl-Sensitive Cut Counting: Endonuclease Digestion Followed by Sequencing: Snipping the regions of interest for sequencing and mapping the sequenced data back to the whole genome can be alternatively used for analyzing the methylation levels. The basic principle is to digest the unmethylated sites of genomic DNA with HpaII and then ligate adapters with digestion site for another enzyme such as EcoP15I or MmeI. This result in many small fragments are around the HpaII site. Further NGS sequencing of these sites and read mapping back to the genome can give an insight into the methylation levels in the genome. Methylation-specific digital karyotyping (Hu, Yao et al., 2006) and Methyl-sensitive cut counting (MSCC) (Ball, Li et al., 2009) are two techniques that follow this principle. Major drawbacks of this method include false positives from about 6% of probes because of cross-reactivity. Other new enzymes SibEnzyme that cut only the methylates sites include BisI, BlsI, GlaI. GluI, KroI, MteI, PcsI, PkrI (Tarasova, Nayakshina et al., 2008). FspEI, MspJI, and LpnPI are type IIS enzymes from New England Biolabs that have the ability to cut outside the recognized site. These can be used to generate snippets of length 32bp around CpG sites that are fully methylated (Cohen-Karni, Xu et al., 2011). While these methods have pros in terms of skipping the bisulfite conversion step, requiring very high quality of DNA for analysis is a huge con.

2.1.2.3. Methylation Status of Specific Genes of Interest

The following techniques can be chosen when a researcher wants to know the methylation levels in a gene of interest. Bisulfite conversion which was already discussed is one of the options. Another include:

- **Bead Array:** Illumina provides VeraCode Methylation assay (http://www.illumina.com/content/dam/illuminamarketing/documents/products/datasheets/datasheet_veracode_methylation.pdf) which works similar to HumanMethylation450 BeadChip array described earlier with an additional option of choosing about 96 or 384 CpG loci.
- **PCR and Sequencing:** The basic protocol here includes bisulfite conversion of DNA followed by amplifying selected regions by PCR with specially designed primers using MethPrimer software (http://www.urogene.org/methprimer) and finally sequencing. Drawbacks of this method include unspecific amplification, primer design, etc.
- **Pyrosequencing:** This method can be used for low throughput projects where a pyrosequencing reaction is performed after PCR amplification of regions of interest (Mahapatra, Klee et al., 2012). Instruments such as PyroMark from Qiagen or Qseq from Bio Molecular Systems are needed for this analysis.
- **Methylation-Specific PCR:** Bisulfite treated DNA is amplified using two different primers for methylated and unmethylated sequences (Herman, Graff et al., 1996). Relative methylation is calculated from the difference in the Ct values from both the reactions. Though accurate, this method can be used only for one or two CpG sites.
- **PCR with High Resolution Melting:** In this technique, an intercalating dye such as SYBR or Eva green bind to PCR amplified double stranded DNA. Based on the difference in temperature, methylated regions are identified. Methylated cytosine has high melting temperature, so the unmethylated strands dissociate early losing the dye. Therefore, strands with the fluorescence are tagged as methylated ones. Though accurate, this method requires pure DNA samples (Wojdacz, Dobrovic et al., 2008).

- **COLD-PCR for the Detection of Unmethylated Islands:** This technique is sensitive and can rare unmethylated sequences can be detected even among large amounts of methylated copies. It is based on the principle that CpG islands in the cancerous tissues are hypomethylated than the normal heavily methylated islands. The protocol includes the bisulfite conversion followed by lowering of denaturation temperature which amplifies only unmethylated fragments (Castellanos-Rizaldos, Milbury et al., 2014).

2.1.3. Techniques for DNA Sequencing

Determining the sequence information of the genome has always been an area of interest since it can reveal a lot about the organism. Sequencing of DNA was first highlighted during the 1970s when two different methods were proposed to identify the sequence information. Later several generations of sequencing technologies have come up that have eventually brought down the associated with cost and time (Goodwin et al., 2016).

2.1.3.1. Sanger Sequencing

It was with the separate works done by Sanger and colleagues and Maxam and Gilbert (Maxam and Gilbert, 1977), focus in the field of science was shifted towards DNA sequencing. Sanger along with his colleagues developed a method which is famously known as Sanger sequencing or chain termination method (Sanger, Nicklen et al., 1977). The sequence of DNA is read by sequence extension and termination where DNA sample is divided into four different reactions. Each reaction mixture consists of a DNA polymerase, three normal dNTPs, a primer and one of the nucleotides modified as ddNTP. Each reaction mixture has one modified ddNTP. These ddNTPs result in the termination of the polymerase reaction due to the lack of 3 oH group that stops the addition of another dNTP molecule to it. After extension, the denatured samples are separated based on their size using gel electrophoresis. Four lanes for each one of the reaction mixtures are used in the gel. DNA sequence is then read directly from the gel. This method is now automated where fluorescent ddNTPs are used for detection and all four-reaction mixtures are run as one single reaction. This method is most widely accepted as toxic chemicals or radioisotopes are not used unlike the chemical degradation method designed by Maxam and Gilbert. Sanger sequencing in fact was used to sequence human genome as a part of the famous Human genome Project which was successfully completed in 2001

(Lander, Linton et al., 2001, Venter, Adams et al., 2001). This method is still more accurate than any of the advanced technologies being used currently but comes with a huge cost burden even for very low throughput.

The need for high throughput sequencing platforms which are more affordable and time saving led to the various advancements in the field of sequencing technologies in terms of capacity, genome complexity, time and cost. Currently, these sequencers are being even used as clinical tools as they give out reads with lengths as long as the whole genome of few organisms at very affordable rates. These advanced technologies are still not comparable with Sanger sequencing in terms of accuracy and read lengths. Therefore, they require a lot of downstream analysis and troubleshooting to achieve a near perfect genome.

High throughput technologies commonly referred as Next Generation Sequencing technologies (NGS) follow a similar paradigm which includes preparation of template, clonal amplification followed by parallel sequencing. They differ in terms of their strategies at every step which therefore differs with each other in terms of quality, quantity, usefulness, etc. These technologies can be categorized broadly into short read sequencing technologies and long read technologies.

2.1.3.2. Short-Read NGS

Template generation is the first step in any of the sequencing technology protocols. Template generation can be achieved by using several strategies such as bead based, solid-state and DNA nanoball generation. (Table 2.66; Figure 2.8). Sequencing after template generation can be done either by sequencing by ligation (SBL) or sequencing by synthesis (SBS). While the former approach uses a fluorophore attached probe sequence is used for hybridization with a template for identification based on the signal emitted, the latter uses a polymerase along with different signal detection strategies such as fluorescence or change in ionic concentrations. Both the approaches make use of a solid surface to clonally amplifying the DNA so that a strong signal can be achieved amidst background noise. Sequencing is also in parallel performed in several lanes or reaction centers to achieve a high throughput.

Table 2.6. Strategies for Template Generation

Template generation methods	Strategy	Platforms using this strategy
Bead based	Bead with immobilized oligonucleotides distributed on a glass surface (Kim, Porreca et al., 2007) or arrayed on a PicoTiterPlate (Leamon, Lee et al., 2003) along with complementary adaptor are used Allows clonal amplification of DNA template on the bead by Emulsion PCR (emPCR)(Dressman, Yan et al., 2003) fig (8a)	454 (Roche), SOLiD (Thermo Fisher), GeneReader (Qiagen), Ion Torrent (Thermo Fisher)
Solid-state (Fedurco, Romieu et al., 2006)	Amplification is performed on a slide which has covalently bound forward and reverses primers. ssDNA binds to the primers with complementary ends Supports Clonal amplification in clusters that are localized, non-overlapping with spatial integrity (Figure 2.8 b,c)	Illumina, SOLiD Wildfire (Thermo Fisher)
DNA nanoball generation (Drmanac, Sparks et al., 2010)	Solution based template enrichment for obtaining a circular template by ligation, circularization, and cleavage using four distinct adaptor regions Generation of billions of DNA nanoballs by rolling circle amplification (RCA) Distribution of nanoball mixture onto a patterned slide surface such that each nanoball has a specific location fig (8 d)	Complete Genomics (BGI)

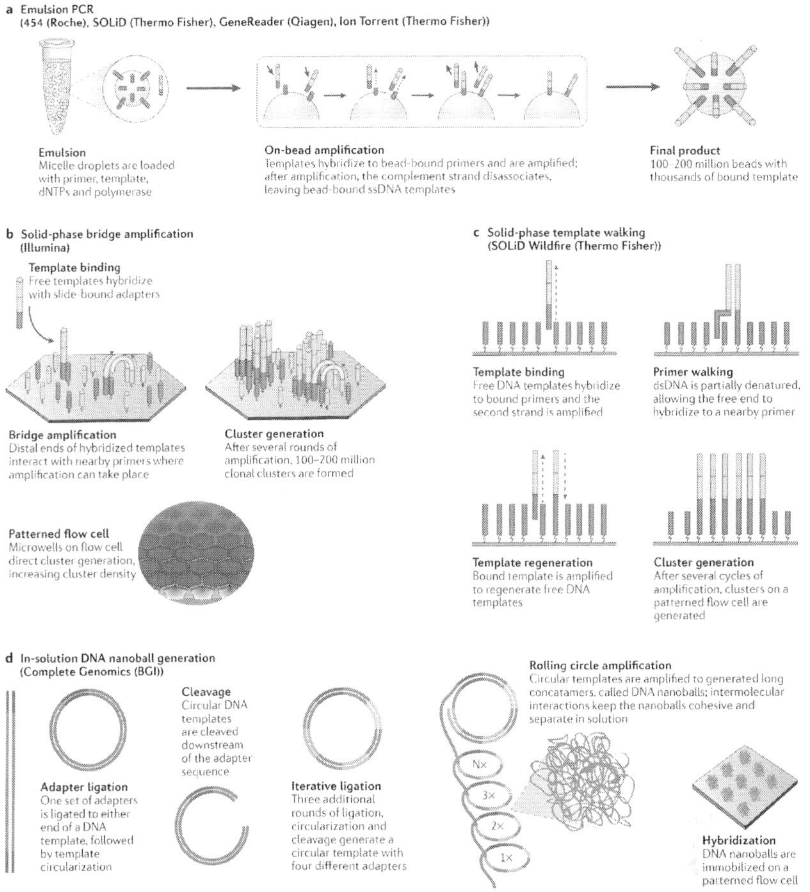

Figure 2.8. Strategies for template generation (a: bead-based, b & c: solid state, d: DNA nanoball generation).

(i) **Sequencing by ligation (SOLiD and Complete Genomics):** Sequencing by ligation approach is performed by forming anchor-probe complexes. Fluorophore labeled probes which encode one or two known bases followed by a series of degenerate or universal bases which are complementary to the template are used along with anchor fragments that are complementary to the adaptor sequence are used. Anchor fragments help in the initiation of ligation and then identification of known bases in the probe help is carried out. This is repeated for several cycles and every time the anchor probe complexes are removed to ensure regeneration of ligation sites for the new cycle. Platforms making use of this approach are SOLiD and Complete Genomics.

SOLiD platform performs sequencing using the following protocol (Figure 2.9 a):

Cluster generation by bead deposition onto a slide.

Fluorophore-labeled two-base-encoded probe ligates to anchor which is complementary to an adaptor.

Identification of first two bases using imaging.

Maintenance of cycle synchronization by capping of unextended strands by unlabeled probes or phosphatase

Cleaving of terminal degenerate bases and the fluorophore so that the next probe can ligate

Repetition of the cycle for ten times so that two out of every five bases are identified.

Removal of ligated probes and repetition of whole cycle with n + 1, n + 2, n + 3 or n + 4 anchors

Complete Genomics either uses combinatorial probe–anchor ligation (cPAL) (Drmanac, Sparks et al., 2010) or combinatorial probe–anchor synthesis (cPAS) to sequence the DNA. cPAL is performed using the following steps (Figure 2.9 b)

Several one-based-encoded probes with a known base at one position and a fluorophore are used in combination with an anchor complementary to the adaptor sequences for hybridization with DNA nanoballs at several locations followed by imaging.

Removal of the entire probe–anchor complexes to make way for new probe–anchor combination.

Every cycle uses a probe pool with known base in the n+1 position and adaptors of variable lengths and chemistries so that sequencing occurs both upstream as well as downstream of the adaptor sequence.

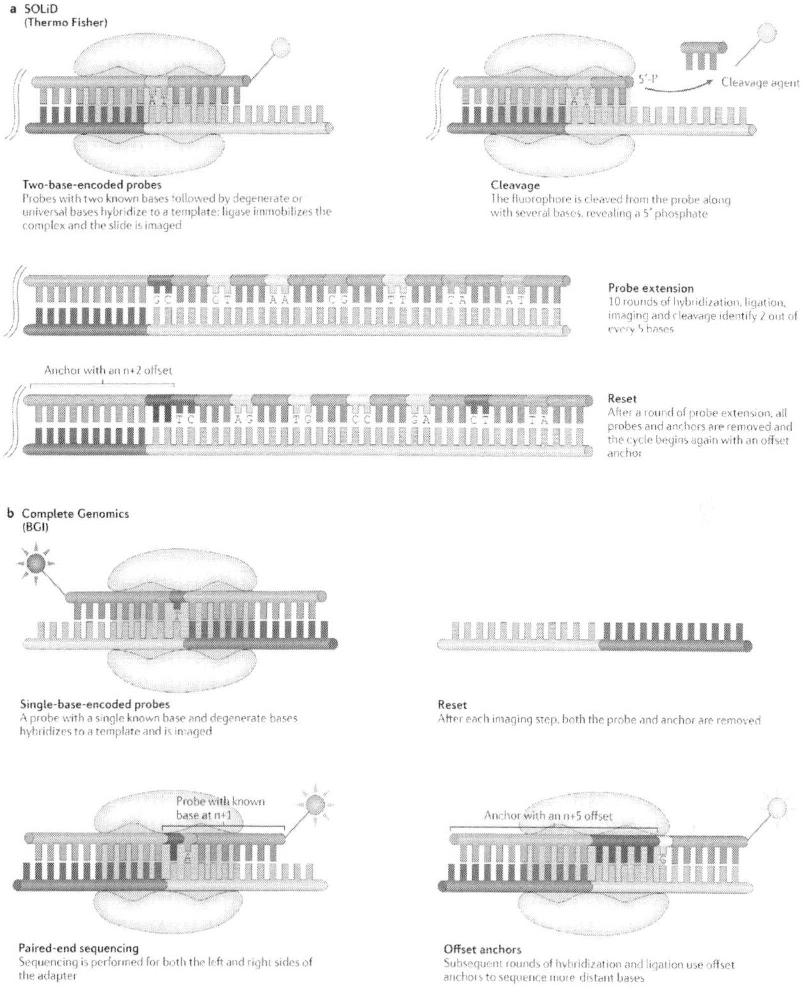

Figure 2.9. Sequencing by ligation (a: SOLiD and b: Complete Genomics).

(ii) Sequencing-by-synthesis

This approach makes use of DNA polymerase and is further classified as cyclic reversible termination (CRT) or as a single-nucleotide addition (SNA) (Metzker, 2010) as shown in Figure 2.10.

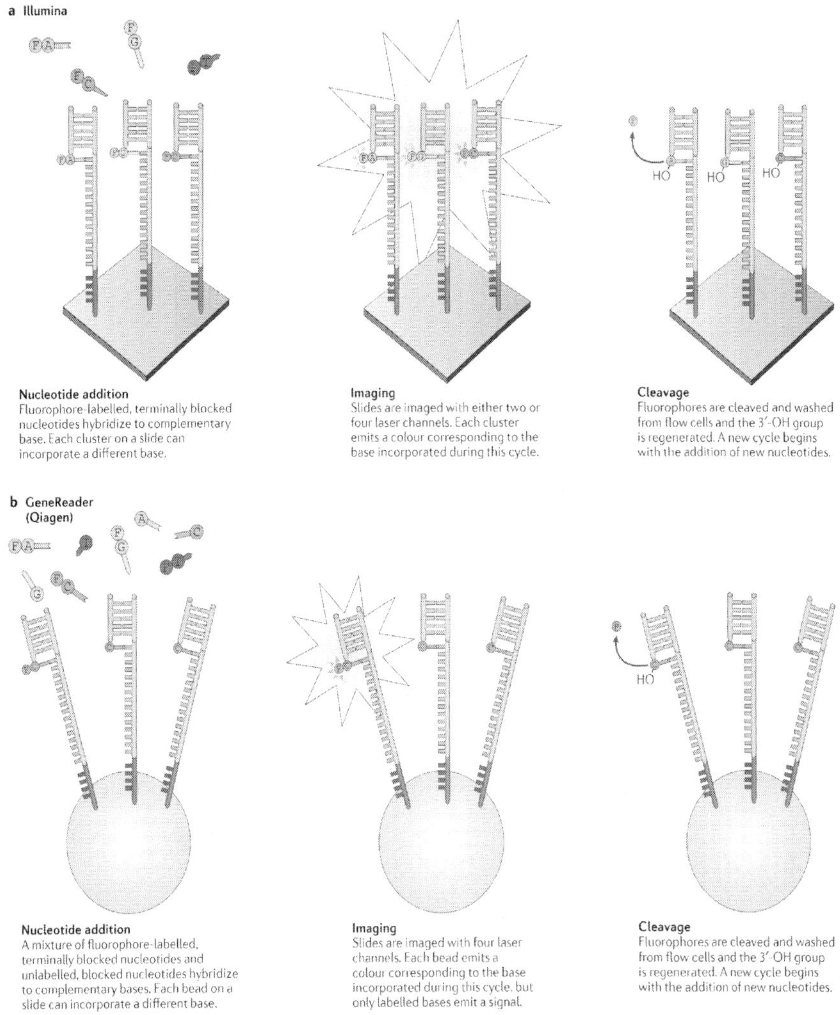

Figure 2.10. Sequencing by synthesis CRT (Illumina, Qiagen) Sequencing by synthesis: CRT (Illumina, Qiagen):

This method uses terminator molecules with 3' OH blocked like in case of Sanger sequencing. The steps involved are

A primer with sequence complementary to the adapter region primes the DNA template and initiates the binding of polymerase

Polymerase then adds the matching labeled and 3′-blocked deoxynucleotides (dNTPs) in every cycle

After every addition, dNTPs that are unbound are removed and then imaging is performed. The labeling on the bound dNTP helps in identifying the base.

To begin the next cycle, fluorophore and blocking groups are removed and the whole process is repeated

One of the key players in this domain, Illumina makes use of this approach. They have a wide range of instruments that differ in terms of throughput, readlengths, etc. Here, total internal reflection fluorescence (TIRF) microscopy is used to identify the dNTP (Figure 2.10a). While most of the systems use four imaging channels to identify the single fluorophore bound to the dNTP, NextSeq and Mini-Seq systems use a two-fluorophore system.

Another platform that uses this approach is Intelligent BioSystems CRT platform which was acquired by Qiagen in 2012 and is now available as GeneReader (Figure 2.10 b). GeneReader performs all the steps starting from sample preparation on a single platform. This platform has two systems, QIAcube sample preparation system and the Qiagen Clinical Insight platform for variant analysis. Its methodology is very much similar to Illumina and only differs in case of incorporation of dNTPs.

Sequencing by synthesis: SNA (454, Ion Torrent)

In this approach, a single nucleotide is responsible for the signal. Therefore, in one cycle only one of the dNTPs is added. This allows skipping of the blocking step. 454 pyrosequencing (Margulies, Egholm et al., 2005) is the first NGS platform to be developed and it makes use of this approach. A PicoTiterPlate with template-bound beads and beads containing an enzyme cocktail is used. Everytime, a dNTP is added to the strand, a bioluminescence signal is emitted. This signal is then detected by charge-coupled device (CCD) camera. (Figure 2.11a).

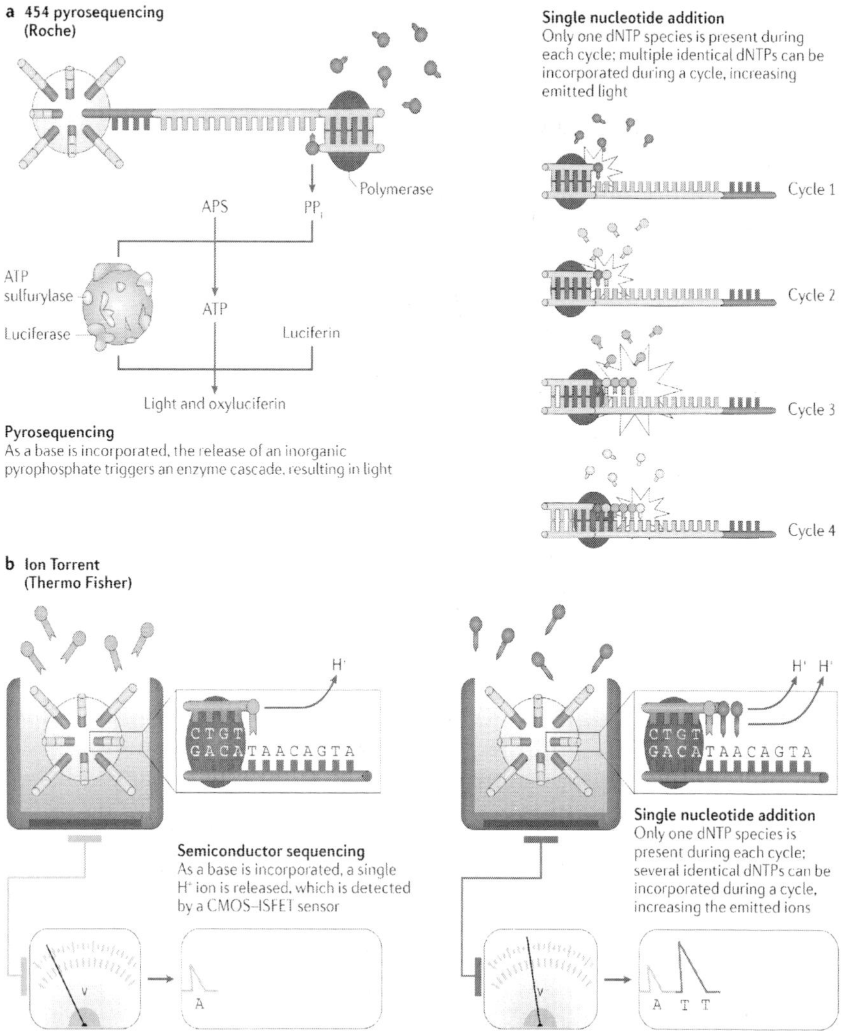

Figure 2.11. Sequencing by synthesis SNA (a: 454, b: Ion Torrent).

Table 2.7. The currently Available Options for Whole Genome Sequencing

Platform	Read length (bp)	Throughput	Reads	Runtime	Error profile	Instrument cost (US$)	Cost per Gb (US$, approx.)
Sequencing by ligation							
SOLiD 5500 Wildfire	50 (SE)	80 Gb	~700 M*	6 d*	≤0.1%, AT bias‡	NA§	$130‡
	75 (SE)	120 Gb					
	50 (SE)*	160 Gb*					
SOLiD 5500xl	50 (SE)	160 Gb	~1.4 B*	10 d*	≤0.1%, AT bias‡	$251,000‡	$70‡
	75 (SE)	240 Gb					
	50 (SE)*	320 Gb*					
BGISEQ-500 FCS[155]	50–100 (SE/PE)*	8–40 Gb*	NA‖	24 h*	≤0.1%, AT bias‡	$250 (REF. 155)	NA‖
BGISEQ-500 FCL[155]	50–100 (SE/PE)*	40–200 Gb*	NA‖	24 h*	≤0.1%, AT bias‡	$250,000 (REF. 155)	NA‖
Sequencing by synthesis: CRT							
Illumina MiniSeq Mid output	150 (SE)*	2.1–2.4 Gb*	14–16 M*	17 h*	<1%, substitution‡	$50,000 (REF. 118)	$200–300 (REF. 118)
Illumina MiniSeq High output	75 (SE)	1.6–1.8 Gb	22–25 M (SE)*	7 h	<1%, substitution‡	$50,000 (REF. 118)	$200–300 (REF. 118)
	75 (PE)	3.3–3.7 Gb	44–50 M (PE)*	13 h			
	150 (PE)*	6.6–7.5 Gb*		24 h*			
Illumina MiSeq v2	36 (SE)	540–610 Mb	12–15 M (SE)	4 h	0.1%, substitution‡	$99,000‡	~$1,000
	25 (PE)	750–850 Mb	24–30 M (PE)*	5.5 h			$996
	150 (PE)	4.5–5.1 Gb		24 h			$212
	250 (PE)*	7.5–8.5 Gb*		39 h*			$142‡
Illumina MiSeq v3	75 (PE)	3.3–3.8 Gb	44–50 M (PE)*	21–56 h*	0.1%, substitution‡	$99,000‡	$250
	300 (PE)*	13.2–15 Gb*					$110‡
Illumina NextSeq 500/550 Mid output	75 (PE)	16–20 Gb	Up to 260 M (PE)*	15 h	<1%, substitution‡	$250‡	$42
	150 (PE)*	32–40 Gb*		26 h*			$40‡
Illumina NextSeq 500/550 High output	75 (SE)	25–30 Gb	400 M (SE)*	11 h	<1%, substitution‡	$250‡	$43
	75 (PE)	50–60 Gb	800 M (PE)*	18 h			$41
	150 (PE)*	100–120 Gb*		29 h*			$33‡
Illumina HiSeq2500 v2 Rapid run	36 (SE)	9–11 Gb	300 M (SE)*	7 h	0.1%, substitution‡	$690‡	$230
	50 (PE)	25–30 Gb	600 M (PE)*	16 h			$90
	100 (PE)	50–60 Gb		27 h			$52
	150 (PE)	75–90 Gb		40 h			$45
	250 (PE)*	125–150 Gb*		60 h*			$40‡
Illumina HiSeq2500 v3	36 (SE)	47–52 Gb	1.5 B (SE)	2 d	0.1%, substitution‡	$690‡	$180
	50 (PE)	135–150 Gb	3 B (PE)*	5.5 d			$78
	100 (PE)*	270–300 Gb		11 d*			$45‡
Illumina HiSeq2500 v4	36 (SE)	64–72 Gb	2 B (SE)	29 h	0.1%, substitution‡	$690‡	$150
	50 (PE)	180–200 Gb	4 B (PE)*	2.5 d			$58
	100 (PE)	360–400 Gb		5 d			$45
	125 (PE)*	450–500 Gb*		6 d*			$30‡
Illumina HiSeq3000/4000	50 (SE)	105–125 Gb	2.5 B (SE)*	1–3.5 d*	0.1%, substitution‡	$740/$900 (REF. 156)	$50
	75 (PE)	325–375 Gb					$31
	150 (PE)*	650–750 Gb*					$22 (REF. 157)

Platform	Read length (bp)	Throughput	Reads	Runtime	Error profile	Instrument cost (US$)	Cost per Gb (US$, approx.)
Sequencing by synthesis: SNA (cont.)							
Illumina HiSeq X	150 (PE)*	800–900 Gb per flow cell*	2.6–3 B (PE)*	<3 d*	0.1%, substitution[‡]	$1,000[‡,§]	$7.0[‡]
Qiagen GeneReader	NA[‖]	12 genes; 1,250 mutations[22]	NA[‖]	Several days[22]	Similar to other SBS systems[22]	NA[‖]	$400–$600 per panel[22]
Sequencing by synthesis: SNA							
454 GS Junior	Up to 600; 400 average (SE, PE)*	35 Mb*	~0.1 M*	10 h*	1%, indel[†]	NA[§]	$40,000[‡]
454 GS Junior+	Up to 1,000; 700 average (SE, PE)*	70 Mb*	~0.1 M*	18 h*	1%, indel[‡]	$108,000[‡]	$19,500[‡]
454 GS FLX Titanium XLR70	Up to 600; 450 mode (SE, PE)*	450 Mb*	~1 M*	10 h*	1%, indel[†]	NA[§]	$15,500[†]
454 GS FLX Titanium XL+	Up to 1,000; 700 mode (SE, PE)*	700 Mb*	~1 M*	23 h*	1%, indel[†]	$450,000[‡]	$9,500[†]
Ion PGM 314	200 (SE)	30–50	400,000–550,000*	23 h	1%, indel[‡]	$49[‡]	$25–3,500[‡]
	400 (SE)	60–100 Mb*		3.7 h*			
Ion PGM 316	200 (SE)	300–500 Mb	2–3 M*	3 h	1%, indel[‡]	$49[‡]	$700–1,000[‡]
	400 (SE)*	600 Mb–1 Gb*		4.9 h*			
Ion PGM 318	200 (SE)	600 Mb–1 Gb	4–5.5 M*	4 h	1%, indel[†]	$49[†]	$450–800[†]
	400 (SE)*	1–2 Gb*		7.3 h*			
Ion Proton	Up to 200 (SE)	Up to 10 Gb*	60–80 M*	2–4 h*	1%, indel[‡]	$224[‡]	$80[‡]
Ion S5 520	200 (SE)	600 Mb–1 Gb	3–5 M*	2.5 h	1%, indel[‡]	$65 (REF. 158)	$2,400*
	400 (SE)*	1.2–2 Gb*		4 h*			$1,200*
Ion S5 530	200 (SE)	3–4 Gb	15–20 M*	2.5 h	1%, indel[‡]	$65 (REF. 158)	$950*
	400 (SE)*	6–8 Gb*		4 h*			$475*
Ion S5 540	200 (SE)*	10–15 Gb*	60–80 M*	2.5 h*	1%, indel[†]	$65 (REF. 158)	$300*
Single-molecule real-time long reads							
Pacific BioSciences RS II	~20 Kb	500 Mb–1 Gb*	~55,000*	4 h*	13% single pass, ≤1% circular consensus read, indel[‡]	$695[‡]	$1,000[‡]
Pacific Biosciences Sequel	8–12 Kb[69]	3.5–7 Gb*	~350,000*	0.5–6 h*	NA[‖]	$350 (REF. 69)	NA[‖]
Oxford Nanopore MK 1 MinION	Up to 200 Kb[159]	Up to 1.5 Gb[159]	>100,000 (REF. 159)	Up to 48 h[160]	~12%, indel[159]	$1,000*	$750*
Oxford Nanopore PromethION	NA[‖]	Up to 4 Tb*	NA[‖]	NA[‖]	NA[‖]	$75*	NA[‖]
Synthetic long reads							
Illumina Synthetic Long-Read	~100 Kb synthetic length*	See HiSeq 2500	See HiSeq 2500	See HiSeq 2500	See HiSeq 2500 (possible barcoding and partitioning errors)	No additional instrument required	~$1,000*
10X Genomics	Up to 100 Kb synthetic length*	See HiSeq 2500	See HiSeq 2500	See HiSeq 2500	See HiSeq 2500 (possible barcoding and partitioning errors)	$75 (REFS 72,161)	See HiSeq 2500 +$500 per sample[161]

Ion torrent is another platform that has adapted this method (Rothberg, Hinz et al., 2011). This was the first platform which does not depend on the optical sensing. Instead of the optical signal, it detects H+ ions that are released when each dNTP is added. Integrated complementary metal-oxide-semiconductor (CMOS) and an ion-sensitive field-effect transistor (ISFET) are used to detect the changes in pH (Figure 2.11 b).

Currently, each of these platforms has several instruments that vary in terms of throughput, cost, and error profile and read structure. Table 2.7 gives an insight into the currently available options for whole genome sequencing.

Long-read sequencing

Complex genomes with highly repetitive regions, structural variations and different copy number of genes require a lot of troubleshooting with short read NGS. Therefore long read sequencing is now favored at least in case of troublesome genomes. With longer read lengths, complex regions are covered within a single read thus avoiding the ambiguity in terms of positions or size. Two approaches namely single-molecule real-time sequencing and synthetic approaches can give long reads (Figure 2.12).

Single-molecule real-time sequencing (PacBio and ONT)

This approach completely avoids clonal amplification which is a mandatory and important step in short read sequencing. Performing the sequencing in cycles is also avoided here. PacBio is the most widely used long read sequencing platform where a specialized flow cell with many thousands of individual picoliter wells with transparent bottoms — zero-mode waveguides are used. A circular template is used so that the same template is sequenced many times. The protocol followed by PacBio is

Fixation of the polymerase to the bottom of the well followed by the progression of DNA strand through ZMV (Levene, Korlach et al., 2003). This allows sequencing of a single molecule instead of sequencing a clone of many DNA molecules.

Addition of a dNTP and visualization by laser and camera system.

Cleaving of flourophore by polymerase before addition of a new dNTP.

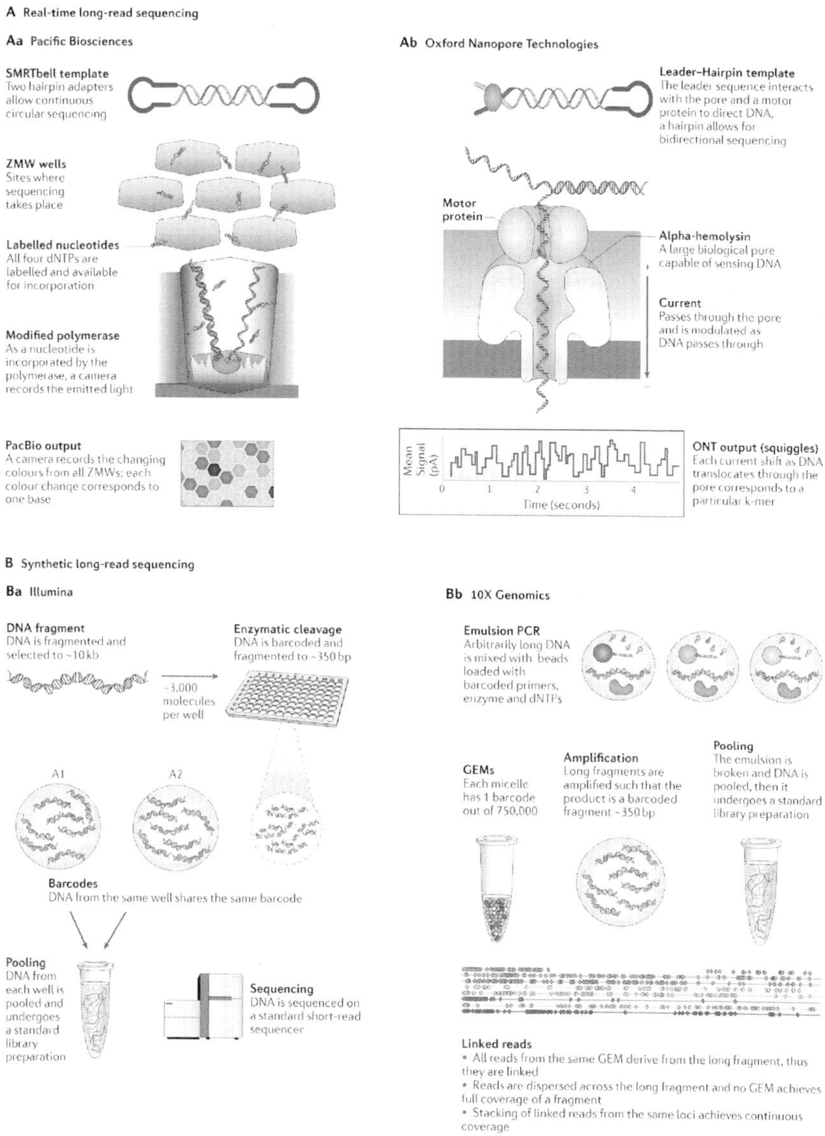

Figures 2.12. Approaches for long read sequencing.

Synthetic long-reads:

Here sequencing of barcoded fragments is performed on the short read sequencers. Re-assemblage of the sequencing information from fragments into a longer fragment is then done using the barcode information. Assembly problems such as breaks (gaps) and unresolvable branch points can be addressed using this technique. Illumina synthetic long-read sequencing platform (Figure 2.12) and the 10X Genomics emulsion-based system are the ones that provide this method of sequencing.

2.2. TRANSCRIPTOME ANALYSIS

The information in the DNA is expressed through transcription. All the transcripts together make up the transcriptome of that organism and their study is known as transcriptomics. Transcriptomics can reveal the total transcripts that were present in the cell at a particular time. This can reveal the regulatory information of the genes. Annotating the unknown proteins can also be done using these studies.

It was around 1990 when transcriptomics evolved (Velculescu, Zhang et al., 1997; Piétu, Mariage-Samson et al., 1999). The very first attempt to capture the human transcriptome of partially which includes ~ 906 transcripts from brain was made during 1991 (Adams, Kelley et al., 1991). This was tailed later by two more reports of two human transcriptomes during 2008 (Pan, Shai et al., 2008) (Sultan, Schulz et al., 2008) followed by many more thousands of such reports (Lappalainen, Sammeth et al., 2013). This vast amount of transcriptome data that is being generated can be attributed to the rapid developments of the new technologies that have greater sensitivity and affordability. One of the first techniques used for transcriptome analysis is Serial Analysis of Gene Expression (SAGE). This was later followed by other contemporary techniques such as microarrays and RNA-Seq.

2.2.1. SAGE

One of the oldest techniques used for capturing a snapshot of the mRNAs in a sample. This technique was later updated into Long SAGE, RL-SAGE, and the most recent SuperSAGE. The main advantage of this technique is that it can help in performing both quantitative as well as analysis of a huge number of transcripts even when there is no information on the gene content (Velculescu, Zhang et al., 1995) (Figure 2.13). This method works in combination with Sanger sequencing. The basic principle of this technique includes two major steps. The first step is representing the mRNAs (cDNAs) in the form of short sequence tags and the second step involves concatenating these tags in the clones for sequence analysis.

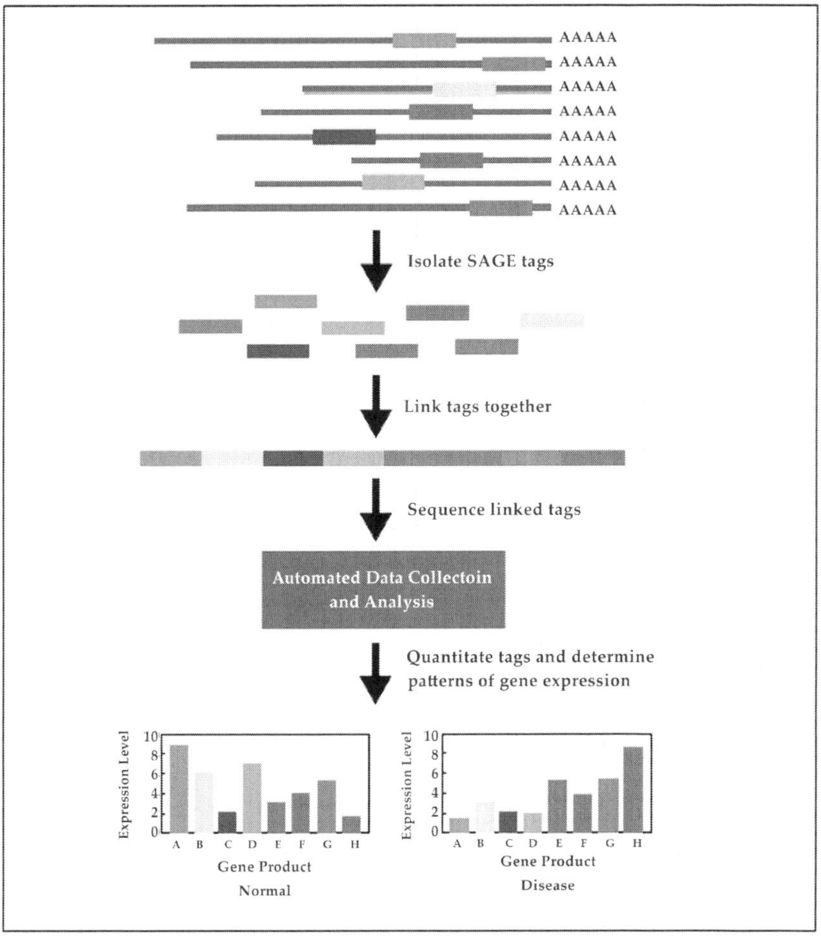

Figure 2.13. SAGE method and tagging.

The principle behind tags representation is that a short unique stretch of 9bp can be used to distinguish 262,144 numbers of transcripts (Yamamoto, Wakatsuki et al., 2001). When all these tags are concatenated, sequencing of these clones can reveal the information in one step (Figure 2.14). The protocol for SAGE involves the following steps

Biotinylated oligo (dT) primer is used to synthesize double stranded cDNA.

A restriction enzyme known as anchoring enzyme is used to cleave the cDNA. A mostly preferred enzyme in this step is NlaIII.

Thus cleaved cDNA is then bound to streptavidin beads and separated into two halves.

Each half of the cDNA is then ligated with two different linkers with enzyme sites for tagging enzymes near the 3'-NlaIII sequence.

These mixtures are then digested with tagging enzymes and the recovery of the released portions is done.

A T4 DNA polymerase is then used to blunt-end the staggered ends of the product.

Now, both the halves are mixed and ligated.

The only mRNA derived terminus is now ligated in a tail-to-tail fashion as the 5' ends of the linkers are blocked by amino groups.

The final product is then amplified using PCR and then an anchoring enzyme is used to cleave the products.

Finally, they are separated by PAGE.

Now, these fragments have flanked ends with NlaIII cohesive terminus and the ligated into concatemers.

These concatemers are then cloned into a vector and sequenced.

LongSAGE and RL-SAGE: These are higher throughput versions of SAGE, while LongSAGE can generate thousands of tags for cDNA library from 20 μg of mRNA (Saha, Sparks et al., 2002), much improved Robust LongSAGE (RL-SAGE) (Gowda, Jantasuriyarat et al., 2004) can complete its cDNA library with just 50 ng mRNA and also required very low number of ditag polymerase chain reactions (PCR).

SuperSAGE: More advanced version of SAGE (Matsumura, Reich et al., 2003) which cuts longer sequence tags of about 26bp from cDNA using type III-endonuclease EcoP15I of phage P1. This has the advantage of accurate allocation to the tag with its respective transcript. Ditags, in this case, is also less biased when compared to that of traditional SAGE ditags. Also, it can provide analysis of millions of tags simultaneously by sequencing using high-throughput sequencing techniques (next-generation sequencing, i.e., pyrosequencing). Gene expression profiles also called as digital gene expression profiling is more accurate and quantitative.

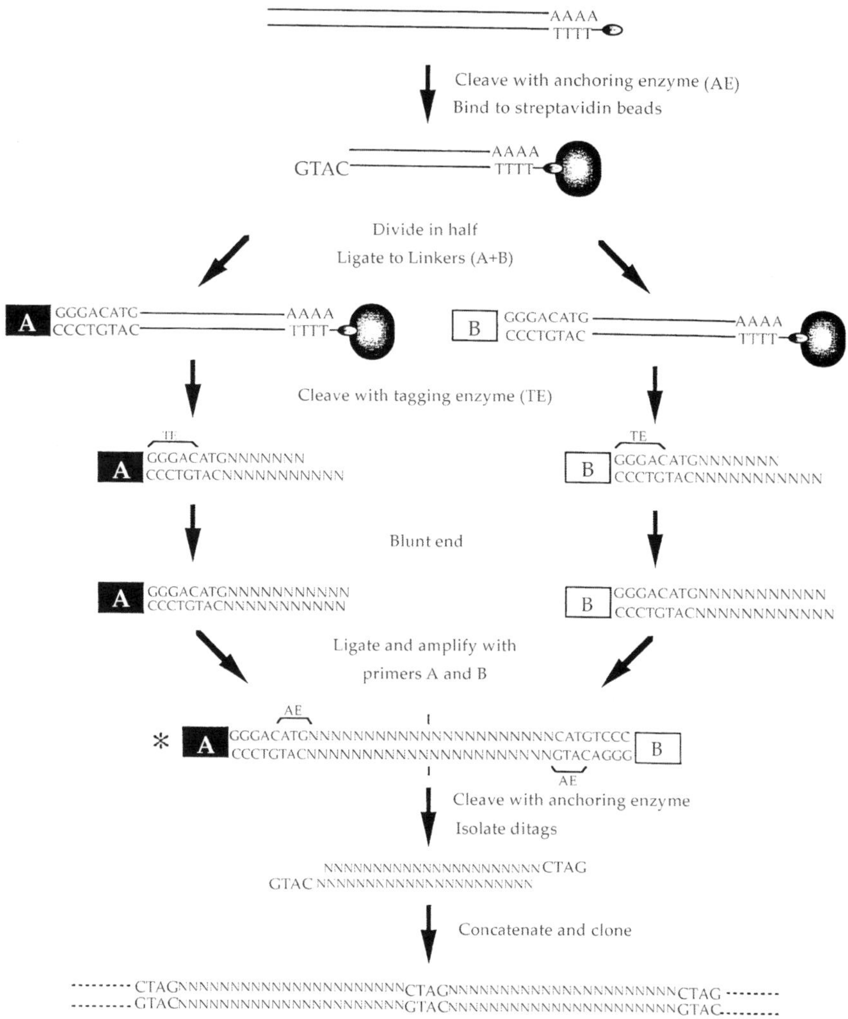

Figure 2.14. Schematic of SAGE procedure (Yamamoto, Wakatsuki et al., 2001).

2.2.2. Microarrays

Microarrays were presented mainly to measure the transcriptional levels of the complete transcriptome in a single experiment. Therefore, it gives an insight into the physiological cell states and expression patterns in certain scenarios such as tumors, estimating the progression of a disease, cell responses and identification of drugs. Presently, microarrays are also used to study SNPs and methylation patterns as discussed in the earlier sections. This is currently considered to be a very powerful tool because of the advancements being made for more than a decade. At present, many different systems of microarrays are available that differ in terms of the solid support being used such as glass or filter surface or the type of DNA fragments used for analysis such as genomic DNA or cDNA or oligonucleotides (Miller and Tang, 2009). The following steps are performed during microarray analysis (Figure 2.15).

- *Probe selection*: The basic principle behind microarrays relies on the ability of the single stranded DNA to form hybrids based on complementarity. In case of transcriptomic analysis, cDNA probes which are unique, sequence-validated and constituting a significant portion of the genome with minimal cross-hybridization to related sequences are used. These probes obtained from cDNA libraries or collections of clones are spotted at precise locations on a solid support which is usually a glass slide or nylon membrane.
- *Target sequences*: The target sequences that hybridize to the probes are the transcripts from the sample of interest. These transcripts are either tagged with a florescence dye or radioisotope after extraction. The whole process can be parallelized for accommodating more than one comparison between the samples by using different fluorophores. Cy3 (green dye) and Cy5 (red dye) are the most commonly used drugs in microarrays.
- *Hybridization*: Now, the labeled targets are hybridized onto the probes for a day or less. After hybridization, the array is washed to remove the unbound target sequences
- *Scanning*: The arrays with probe-target sequences are scanned to identify the fluorescence at each s-pot. These scanned images are then analyzed to determine the expressions of the transcripts.

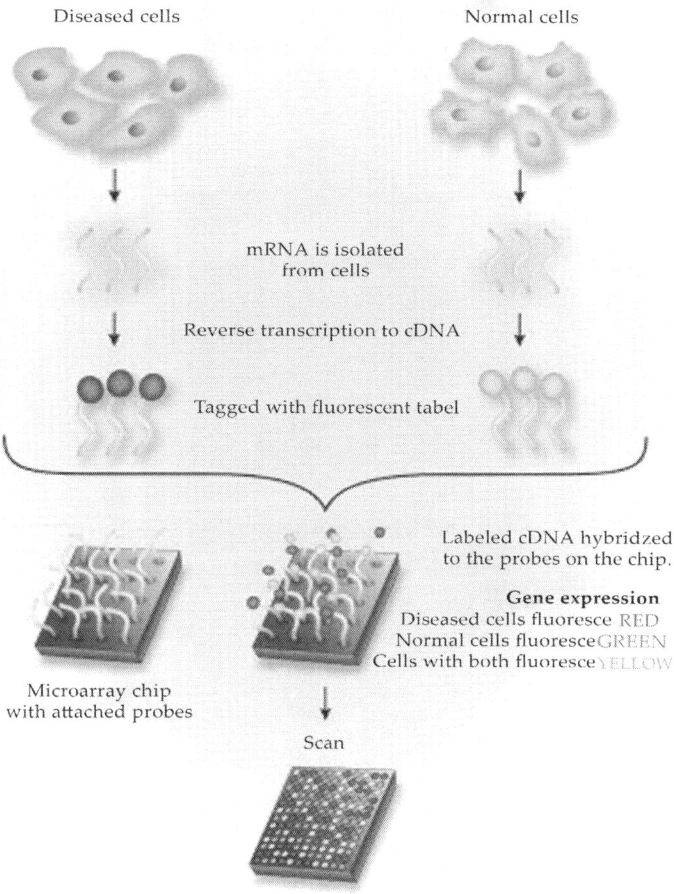

Figure 2.15. Microarray step-step process description (Villaseñor-Park and Ortega-Loayza, 2013).

Though microarrays are advantageous in terms of generating high throughput data, they have few disadvantages as well. Discerning expression differences from experimental design and secondary effects are not really possible with microarrays. In addition, other factors such as the purity of samples, experimental design, and quality of microarray design and data analysis decide the quality and type of the final output data from these microarray experiments. In many cases, the results have to be double checked with quantitative reverse-transcriptase PCR, RNase protection assays, and northern blot analysis (Villaseñor-Park and Ortega-Loayza, 2013).

2.2.3. RNA-Seq

RNA-Seq is an advanced technique used for transcriptomics analysis which came into existence with the advent of next generation sequencing technologies. Before sequencing of RNA transcripts, a lot of pre-processing is required. There are several RNA-Seq methods available that differ in their protocols for transcript enrichment, fragmentation, and amplification. Both single and paired-end sequencing can be performed. The basic protocol of a RNA-Seq experiment involves the following steps (Figure 2.16).

- *Enrichment of certain class of RNA transcripts*: This can be done by removing other abundant RNAs to increase the sensitivity. Separation of mRNA molecules can be done using the probes that are complementary to the poly-A tails. Gel electrophoresis can be used to purify microRNA's. Using taxon-specific rRNA probes can help in depleting the abundant rRNAs.
- *Fragmentation*: A part of sequencing library construction (Knierim, Lucke et al., 2011), fragmentation is usually needed in case of mRNAs as they are longer than the average read lengths given out by the sequencers. There are several ways by which fragmentation can be done; this includes chemical hydrolysis, sonication, or simultaneous fragmentation and tagging of cDNA by transposase enzymes.
- *Amplification*: This step is needed to enrich the sample RNAs and this is usually done by PCR.
- *Sequencing*: Sequencing of the amplified molecules can be done either in one direction (single end sequencing) or in opposite directions (Paired-end sequencing). While single end sequencing is cheaper and less time taking, paired end sequencing provides more rich and deep data for analysis that can be used for better alignments/assemblies. Paired end sequencing is mainly helpful in discovering isoforms (Wang, Gerstein et al., 2009). To preserve the strand information, strand specific sequencing can be performed (Levin, Yassour et al., 2010).

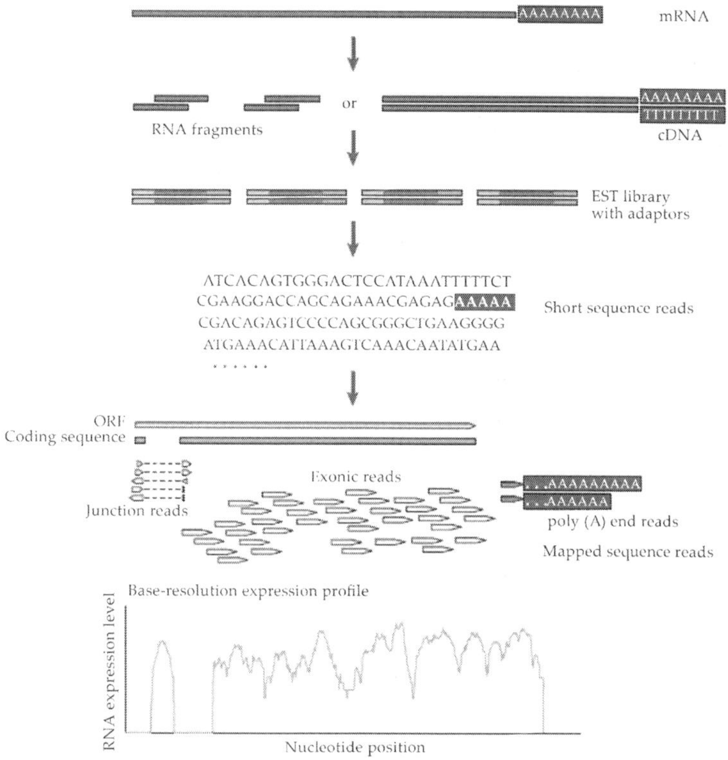

Figure 2.16. RNA-seq experiment (Wang, Gerstein et al., 2009).

Currently available platforms that offer RNA-Seq are listed out in the table 2.8. All these platforms provide same machines for sequencing genomic and cDNA molecules. Therefore, it is mandatory to copy the RNA to cDNA molecules before sequencing. The most advanced and state of art sequencing platform is expected to be nanopore sequencing that can sequence the RNA molecule directly. This is more advantageous in detecting the modified bases that are usually not identified from cDNA sequencing. Amplification steps are also eliminated in nanopore sequencing.

Table 2.8. Sequencing Technology Platforms used for RNA-Seq (Liu, Li et al., 2012) (Quail, Smith et al., 2012)

Platform	Commercial release	Typical read length	Maximum throughput per run	Single read accuracy	RNA-Seq runs deposited in the NCBI SRA (Oct 2016)
454 Life Sciences	2005	700 bp	0.7 Gbp	99.9%	3548
Illumina	2006	50-300 bp	900 Gbp	99.9%	362903
SOLiD	2008	50 bp	320 Gbp	99.9%	7032
Ion Torrent	2010	400 bp	30 Gbp	98%	1953
PacBio	2011	10,000 bp	2 Gbp	87%	160

Advantages of RNA-Seq over other transcriptomic technologies: The major advantages of this technology over others are listed out in the Table 2.9.

Table 2.9. Advantages of RNA-Seq Over Others (Wang, Fan et al., 1998)

Technology	Microarray	EST sequencing	RnA-seq
Technology specifications			
Principle	Hybridization	Sanger sequencing	High-throughput sequencing
Resolution	From several to 100 bp	Single base	Single base
Throughput	High	Low	High
Reliance on genomic sequence	Yes	No	in some cases
Background noise	High	Low	Low
Application			
Simultaneously map transcribed regions and gene expression	Yes	Limited for gene expression	Yes
Dynamic range to quantify gene expression level	Up to a few-hundredfold	Not practical	>8,000-fold
Ability to distinguish different isoforms	Limited	Yes	Yes
Ability to distinguish allelic expression	Limited	Yes	Yes
Practical issues			
Required amount of RNA	High	High	Low
cost for mapping transcriptomes of large genomes	High	High	Relatively Low

2.2.3.1. Limitations of RNA-Seq

A number of reads obtained from the sample play a major role in defining the output data. Sensitivity and accuracy of the experiments are decided based on the coverage of reads. Also, this differs from species to species. Also, the cost involved here is much greater when compared to other techniques. Like other NGS technologies, RNA-Seq also faces downstream bioinformatics challenges (Wang, Gerstein et al., 2009).

2.3. PROTEOMIC ANALYSIS

Proteomics is the study of the entire proteome of a cell or organism which can give an insight into the profiles of the proteome, gene expression level comparisons among the proteins, identification of posttranslational modifications and interactions between proteins (Wilkins, Sanchez et al., 1996). A human genome with an approximate of 26,000–31,000 coding genes harbors more than a million proteins (Wilkins, Gasteiger et al., 1999, Baltimore, 2001, Godovac-Zimmermann and Brown, 2001). Therefore to gain the functional information of any cell or organism, it is necessary to analyze its proteome (Alaoui-Jamali and Xu, 2006). Technologies for studying proteomics are now more advanced when compared to the 2D-gel electrophoresis days. Table 2.10 lists out few technologies that have been in use for studying the proteome.

Table 2.10. Technologies of Proteomics (Chandramouli and Qian, 2009)

Technology	Application	Strengths	Limitations
2DE	Protein separation Quantitative expression profiling	Relative quantitative PTM information	Poor separation of acidic, basis, hydrophobic and low abundant proteins
DIGE	Protein separation Quantitative expression profiling	Relative quantitative PTM information High sensitivity Reduction of intergel variability	Proteins without lysine cannot be labeled Requires special equipment for visualization fluorophores are very expensive
ICAT	Chemical isotope labeling for quantitative proteomics	Sensitive and reproducible Detect peptides with low expression levels	Proteins without cysteine residues and acidic proteins are not detected
SILAC	Direct isotope labeling of cells Differential expression pattern	Degree of labeling is very high Quantitation is straightforward	SILAC labeling of tissue samples is not possible

iTRAQ	Isobaric tagging of peptides	Multiplex several samples Relative quantification High-throughput	Increases sample complexity Require fractionation of peptides before MS
MUDPIT	Identification of protein-protein interactions Deconvolve complex sets of proteins	High separation Large protein complexes identification	Not quantitative Difficulty in analyzing the huge data set Difficult to identify isoforms
Protein array	Quantitate specific proteins used in diagnostics (biomarkers or antibody detection) and discovery research	High-throughput Highly sensitive Low sample consumption	Limited protein production Poor expression methods Availability of the antibodies Accessing very large numbers of affinity reagents
Mass spectrometry	Primary tool for protein identification and characterization	High sensitivity and specificity High-throughput Qualitative and quantitative PTM information	No individual method to identify all proteins. Not sensitive enough to identify minor or weak spots. MALDI and ESI do not favor identification of hydrophobic peptides and basic peptides
Bioinformatics	Analysis of qualitative and quantitative proteomic data	Functional analysis, data mining, and knowledge discovery from Mass spectrometric data	No integrated pipeline for processing and analysis of complex data Search engines do not yield identical results

Protein microarrays and Mass Spectrometry which are the current advanced technologies will be described in brief.

2.3.1. Protein Microarrays

Around the 1980s, DNA or mRNA-derived cDNAs were used for expression analysis on lab-on-chip platforms (Ekins, 1989, MacBeath, 2002). After three decades of advancements, miniature versions of these are commercially available as microarrays (Beaudet and Belmont, 2008, Shiu and Borevitz, 2008). These help in analyzing the interactions and aberrant activities of proteins that plays a major role in diseases such as cancer. Since proteins are more labile than DNA, protein microarray standardization was more challenging than the DNA microarrays (Talapatra, Rouse et al., 2002). Advancements were made to tackle the labile nature of proteins along with other practical problems such as its short shelf life, lack of universal binding partners or amplification modalities (Sasakura, Kanda et al., 2004; Ramachandran, Hainsworth et al., 2006; Syed, Gupta et al., 2015).

The first ever protein microarray was based on sandwich immunoassay platforms (Silzel, Cercek et al., 1998). Antibody arrays were used conventionally where antigens are captured and detected. The most commonly performed antibody based microarray is based on a two-color method where two labels Cy5 and Cy3 are used to label the sample mixture and then used as a probe on to an array with antibodies.

Advancements were further made to these antibody arrays which led to the availability of a broader class of arrays with antibodies, aptamers, etc. (Figure 2.17) (Phizicky, Bastiaens et al., 2003). To detect protein-protein interactions for discovering biomarkers, arrays spotted with purified antigens are used which are again based on immunoassay principle. Different kinds of protein arrays that are currently available are listed out in the Table 2.11.

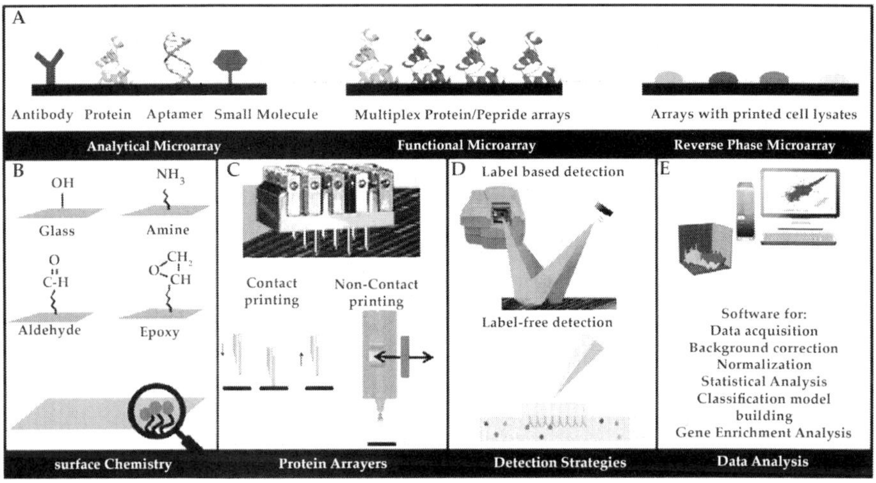

Figure 2.17. Schematic representation of core components of protein micro-array platform A) Commonly used proteins array platforms. Analytical arrays have capture molecules printed in multiplex formats to check for binding partners. Functional arrays have multiple proteins or peptides arrayed on a single chip to screen biospecimens for diagnostics, expression or profiling. RPPA have multiple cell lysates spotted on a single chip to detect the presence of specific proteins/biomarkers in a sample. (B–D): Commonly used surface chemistry, protein arrays, and detection and data analysis platforms, respectively (Table 2.11). (Gupta, Manubhai et al., 2016).

Table 2.11. Commercially Available Protein Array Chips (Gupta, Manubhai et al., 2016)

Type	Product name	Company	Applications
Antibody array	Plasma scan 380 antibody arrays	Arrayit	Biomarker discovery
	Apoptosis, cancer biomarker, AKT pathway, phospho antibody arrays	FULL MOON BioSystems	Protein expression profiling, biomarker identification, phosphorylation studies, disease versus healthy comparisons
	Cell cycle, angiogenesis, AKT-phospho arrays	Abnova	NA
	Human angiogenesis array C1000, cytokine array Q2	Ray Biotech	Detect relative protein expression/biomarker discovery
	Human cytokine array	Spring Bioscience	NA
	BD Clontech antibody array	BD Biosciences	Detects tissue, cells, and different body fluid proteome
	Pathway and phosphorylation antibody microarrays	Eurogentec	Biomarker detection, protein expression profiling
	Apoptosis, cell cycle, and signal transduction arrays	Hypromatrix	Protein–protein interaction, protein phosphorylation
	Panorama antibody arrays	Sigma Aldrich	Protein expression profiling from serum, tissue, and cell
Peptide array	PEPperCHIP MERS-CoV Proteome Microarray	B-bridge international	Epitope mapping and antigen discovery
	Pamchip 4	PamGene	NA

Protein array	HuProt arrays	CDI Laboratories	Human sera based autoantibody screening, specificity of antibody, cross-reaction detection using proteome-wide antibody screening
	HuProt v2 human proteome microarrays	CDI Laboratories	Autoantibody screening, antibody binding specificity profiling, protein–protein interactions, protein–small molecular interactions, protein-DNA/RNA interactions
	TB array *Mycobacterium tuberculosis* proteome microarray	CDI Laboratories	Protein–protein interactions, small molecule–protein binding, serum profiling (patient/population)
	VirD array (functional membrane protein virion display microarray)	-	NA
	ProtoArray human protein microarray	Thermo Fisher Scientific	Autoantibody screening (immune response), protein–protein interaction, small molecule profiling, enzyme substrate, and antibody substrate profiling
	ProtoPlex immune response assay featuring Luminex xMAP technology	-	Helps in screening and validation of autoantibody markers
	Yummy! dietary wellness and environmental allergy microarray tests and testing services	Arrayit	Screen environmental, food, and herbal allergens
	AllerSpot companion animal allergy testing service	-	Deciding suitable pet food, identifying the environmental allergens, and selecting proper allergic medication
	Pathogen antigen arrays	-	To detect antibodies against *M. tuberculosis, P. falciparum, B. pseudomallei, S. typhi, S. aureus*
	Immunome protein arrays	Sengenics	Biomarker discovery for cancer and autoimmune disease; global immune response to viral, fungal, and microbial; global-level immune expression drug responses in clinical trials
	HaloLink protein array systems	Promega	Biomarker discovery, protein–protein, protein–drug, protein–nucleic acid interaction
	Pep chip protease profiling	Pepscan	NA

2.3.2. Mass Spectrometry

Mass spectrometry (MS) is one of the widely used techniques in proteomics where information about the identity of the protein, its structure, and chemical modifications are given out by their mass. A mass spectrometer has three components; source, analyzer, and detector. While a source generates ions from the charged particles such as proteins, analyzer separates these ions based on their mass. Finally, a detector detects the charged particles and gives out a measurable signal after amplifying it.

John Fenn and Koichi Tanaka have developed two techniques namely electrospray and MALDI (matrix-assisted laser desorption ionization mass spectrometry. These two techniques aid in the vaporization as well as ionization of proteins and peptides which are comparatively large and non-volatile, unlike earlier MS techniques that were only limited to volatile and low molecular weight molecules. Later, several developments were to make them more usable in proteomics.

- *Source – Electrospray ionization*: A high voltage needle at about ~2-5kV is used for passing the sample solution through for spraying of analyte containing charged droplets. High charged species are obtained once the solvent absorbs and an explosion occurs as seen in Figure 2.18 ESI uses multiple charged analytes. Mass spectrometers separate ions based on charge to mass ratio (m/z) and for analysing higher charge molecules above their mass range, this method can be used.

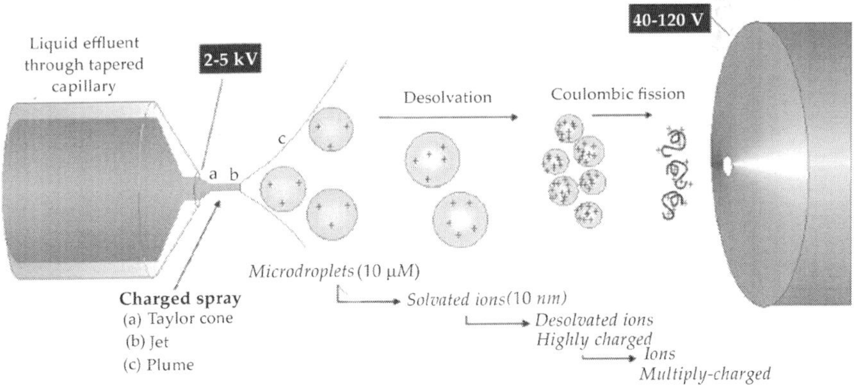

Figure 2.18. Electrospray ionization (Parker, Warren et al., 2010).

- *Source-MALDI:* matrix-assisted laser desorption/ionization is another method which is used for ionization similar to ESI. It vaporizes and ionizes the analyte using laser and chemical compounds that can absorb UV (Figure 2.19). The main drawback of ESI is that it can handle proteins with masses up to 65-70 kDa while MALDI can go up to 250kDa. And with advanced analyzers and detectors, it can still go up. MALDI is also not sensitive to salts and detergents and gives out high throughput, unlike ESI.

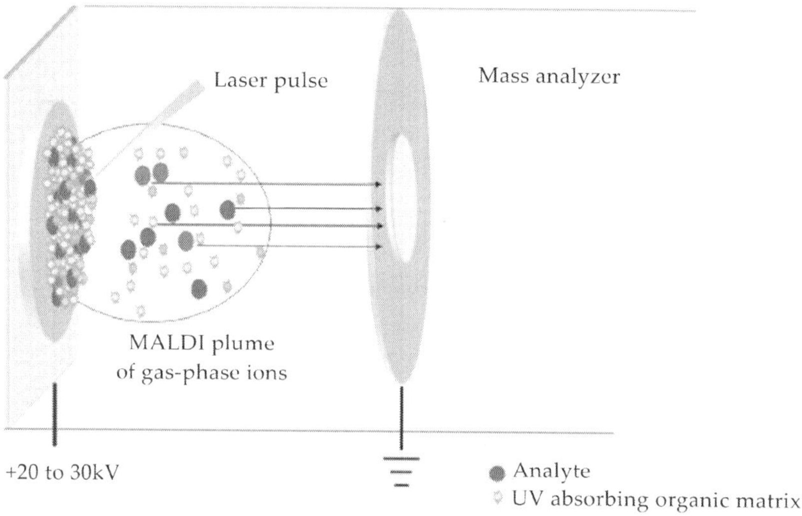

Figure 2.19. MALDI ionization (Parker, Warren et al., 2010).

- *Mass analyzer- Quadrupole Mass Filters*: Earlier mass analyzers could only analyze molecules of lesser masses but with advancements and developments in analyzers also lead to the introduction few state of art techniques such as Quadrupole Mass Filters and Time-of-Flight Analyzers. In case of Quadrupole Mass Filters, only ions with certain m/z ratios are only passed through by using four parallel rods with RC and DC voltage combinations (Figure 2.20).
- *Mass analyzer-Time of flight analyzers:* Here, the time taken by the ion to reach the detector in field free vacuum is calculated to give an indirect indication of m/z ratio. At the zero time, packets of ions are released in to the analyzer from the source and the signal is generated at time t. Now the flight time of the ion is proportional to its mass which means that heavier ones travel

slower than their light weight counterparts. Linear TOF designs were earlier used (Figure 2.20) where a linear path is taken up by the ions for their journey towards the detector. Resolution is sometimes reduced in case of ions that form a plume in the vacuum because of the different acceleration voltages. To overcome this issue, modifications such as delayed extraction of ions from the source as well as adding up reflectron lenses to the original linear design were put to use.

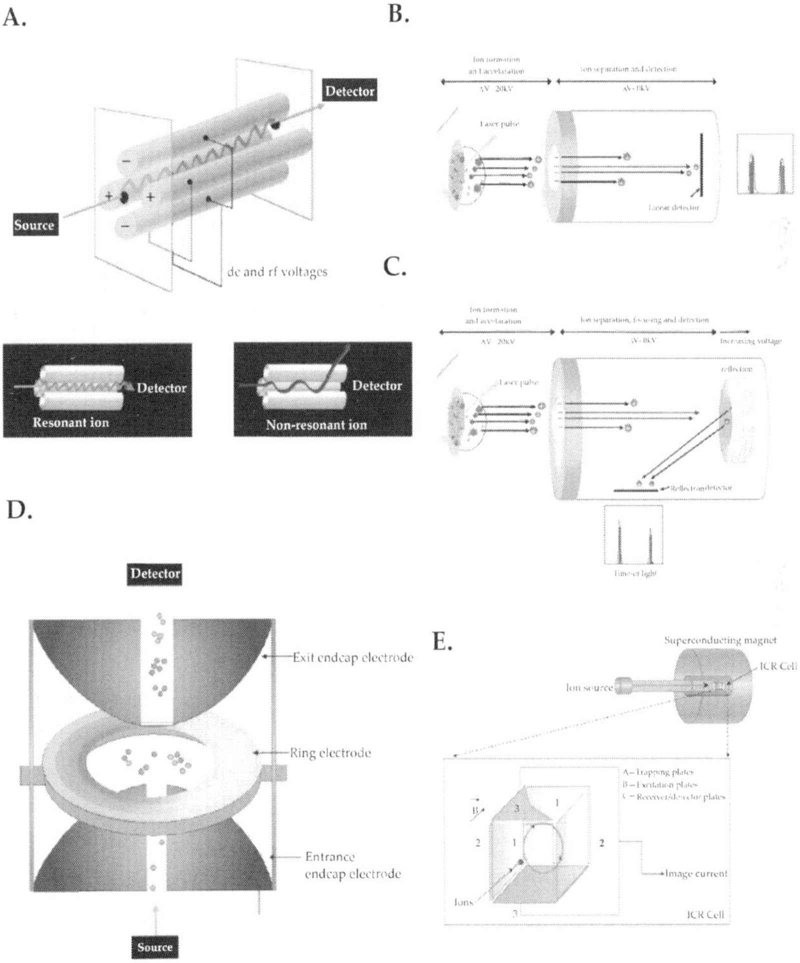

Figure 2.20. Different Analyzers used in MS. (A) quadrupole mass analyzer (B) TOF analyzers – linear (C) TOF analyzers – reflection configuration (D) ion trap (E) FT-ICR (Parker, Warren et al., 2010).

- *Mass analyzer-Ion traps:* This is a modification to quadrupole mass analyzers, where only three instead of four iron rods are used. Two electrodes let the electrons to flow selectively from analyzer to detector while the third one known as ring electrode traps the ions.
- *Mass analyzer- Fourier Transform MS:* This analyzer is rarely used in proteomics though it has the highest resolution of mass. FT-ICR and the Orbitrap are the two versions of this analyzer. In case of former, a pair of plates is arranged in the form of cubes and surrounded by strong magnetic field (Figure 2.20E). This arrangement is called as ICR cell which ICR-MS requires ion trapping, excitation, detection, and data transformation. While in case of Orbitrap, an inner axial electrode is used to trap the ions by increasing the voltage as they enter. This results in a spiral which ultimately stops the increase in the voltage by moving into the center of the axis. Finally, a ring of ions at different frequencies are formed which are again converted to ion masses by Fourier transformation of the ion current.
- *Mass analyzer- Ion-Mobility MS:* This technique uses collisional cross sections of the ions to separate them.
- *Detector:* The final component is a detector which is used to strike the ions that have travelled from the source. This result in the release of electrons from the surface of the detector and the released electrons, again and again, strike the surface resulting in a final stream of electrons that ultimately strike the cathode (Figure 2.21a). For detecting negative ions, a conversion diode is used at the entrance which changes the polarity of the ions. When these ions strike the diode, a positive ion is produced initiating the cascade of electrons similar to the traditional scenario. Both the detection modes have the same voltage. The newly advanced detectors have many multipliers arranged in an array fashion instead of one. These are known as multichannel plate (MCP) detectors (Figure 2.21 b).

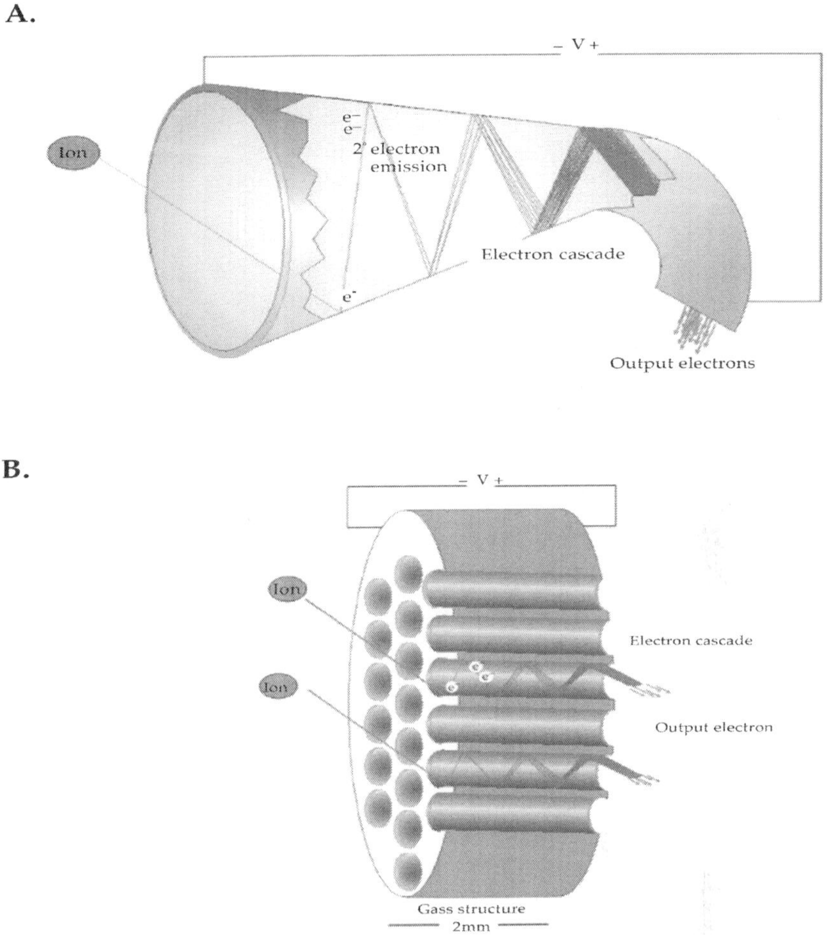

Figure 2.21. Detectors (A) Continuous dynode detector (B) multichannel plate detector (Parker, Warren et al., 2010).

REFERENCES

1. Adams, M. D., Kelley, J. M., Gocayne, J. D., Dubnick, M., Polymeropoulos, M. H., Xiao, H., Merril, C. R., Wu, A., Olde, B., & Moreno, R. F. (1991). "Complementary DNA sequencing: expressed sequence tags and human genome project." *Science 252*(5013), 1651–1656.
2. Alaoui-Jamali, M.A., & Xu, Y. J.(2006). "Proteomic technology for biomarker profiling in cancer: an update." *J Zhejiang Univ Sci B 7*(6), 411–420.
3. Arakawa, Y., Watanabe, M., Inoue, N., Sarumaru, M., Hidaka, Y., & Iwatani, Y. (2012). "Association of polymorphisms in DNMT1, DNMT3A, DNMT3B, MTHFR and MTRR genes with global DNA methylation levels and prognosis of autoimmune thyroid disease." *Clin Exp Immunol 170*(2), 194–201.
4. Aung, H. T., Harrison, D. K., Findlay, I., Mattick, J. S., Martin, N. G., & Carroll, B. J. (2010). "Stringent programming of DNA methylation in humans." *Twin Res Hum Genet 13*(5), 405–411.
5. Ball, M. P., Li, J. B., Gao, Y., Lee, J. H., LeProust, E. M., Park, I. H., Xie, B., Daley, G. Q., & Church, G. M. (2009). "Targeted and genome-scale strategies reveal gene-body methylation signatures in human cells." *Nat Biotechnol 27*(4), 361–368.
6. Baltimore, D. (2001). "Our genome unveiled." *Nature 409*(6822), 814–816.
7. Beaudet, A. L., &, Belmont, J. W. (2008). "Array-based DNA diagnostics: let the revolution begin." *Annu Rev Med 59*, 113–129.
8. Bell, P. A., Chaturvedi, S., Gelfand, C. A., Huang, C. Y., Kochersperger, M., Kopla, R., Modica, F., Pohl, M., Varde, S., Zhao, R., Zhao, X., Boyce-Jacino, M. T., & Yassen, A. (2002). "SNPstream UHT: ultra-high throughput SNP genotyping for pharmacogenomics and drug discovery." *Biotechniques Suppl*, 70–72, 74, 76–77.
9. Bortolin, S., Black, M., Modi, H., Boszko, I., Kobler, D., Fieldhouse, D., Lopes, E., Lacroix, J. M., Grimwood, R., Wells, P., Janeczko, R.., & Zastawny, R..(2004). "Analytical validation of the tag-it high-throughput microsphere-based universal array genotyping platform: application to the multiplex detection of a panel of thrombophilia-associated single-nucleotide polymorphisms." *Clin Chem 50*(11), 2028–2036.

10. Botstein, D., White, R. L., Skolnick, M., & Davis, R. W. (1980). "Construction of a genetic linkage map in man using restriction fragment length polymorphisms." *Am J Hum Genet 32*(3), 314–331.

11. Braun, A., Little, D. P., & Koster, H. (1997). "Detecting CFTR gene mutations by using primer oligo base extension and mass spectrometry." *Clin Chem 43*(7), 1151–1158.

12. Brookes, A. J. (1999). "The essence of SNPs." *Gene 234*(2), 177–186.

13. Bujko, M., Musialik, E., Olbromski, R., Przestrzelska, M., Libura, M., Pastwińska, A., Juszczyński, P., Zwierzchowski, L., Baranowski, P., & Siedlecki, J. A. (2014). "Repetitive genomic elements and overall DNA methylation changes in acute myeloid and childhood B-cell lymphoblastic leukemia patients." *Int J Hematol 100*(1), 79–87.

14. Calvo, X., Nomdedeu, M., Navarro, A., Tejero, R., Costa, D., Muñoz, C., Pereira, A., Peña, O., Risueño, R. M., Monzó, M., Esteve, J., & Nomdedeu, B. (2014). "High levels of global DNA methylation are an independent adverse prognostic factor in a series of 90 patients with de novo myelodysplastic syndrome." *Leuk Res 38*(8), 874–881.

15. Cashman, J. R., Zhang, J., Leushner, J., & Braun, A. (2001). "Population distribution of human flavin-containing monooxygenase form 3: gene polymorphisms." *Drug Metab Dispos 29*(12), 1629–1637.

16. Castellanos-Rizaldos, E., Milbury, C. A., Karatza, E., Chen, C. C., Makrigiorgos, G. M., & Merewood, A. (2014). "COLD-PCR amplification of bisulfite-converted DNA allows the enrichment and sequencing of rare un-methylated genomic regions." *PLoS One 9*(4), e94103.

17. Chandramouli, K., & Qian, P. Y.(2009). "Proteomics: challenges, techniques and possibilities to overcome biological sample complexity." *Hum Genomics Proteomics 2009.*

18. Cohen-Karni, D., Xu, D., Apone, L., Fomenkov, A., Sun, Z., Davis, P. J., Kinney, S. R., Yamada-Mabuchi, M., Xu, S. Y., Davis, T., Pradhan, S., Roberts, R. J., & Zheng, Y. (2011). "The MspJI family of modification-dependent restriction endonucleases for epigenetic studies." *Proc Natl Acad Sci U S A 108*(27), 11040–11045.

19. Consortium, I. H. (2005). "A haplotype map of the human genome." Nature 437(7063), 1299–1320.

20. Conti, A., Kobets, T., Escudero-Lourdes, C., Montgomery, B., Tryndyak, V., Beland, F. A., Doerge, D. R., & Pogribny, I. P. (2014).

"Dose- and time-dependent epigenetic changes in the livers of Fisher 344 rats exposed to furan." *Toxicol Sci 139*(2), 371–380.

21. Dressman, D., Yan, H., Traverso, G., Kinzler, K. W., & Vogelstein, B. (2003). "Transforming single DNA molecules into fluorescent magnetic particles for detection and enumeration of genetic variations." *Proc Natl Acad Sci U S A 100*(15), 8817–8822.
22. Drmanac, R., Sparks, A. B., Callow, M. J., Halpern, A. L., Burns, N. L., Kermani, B. G., et. al. (2010). "Human genome sequencing using unchained base reads on self-assembling DNA nanoarrays." *Science 327*(5961), 78–81.
23. Ekins, R. P. (1989). "Multi-analyte immunoassay." *J Pharm Biomed Anal 7*(2), 155–168.
24. Faham, M., Baharloo, S., Tomitaka, S., DeYoung, J., & Freimer, N. B. (2001). "Mismatch repair detection (MRD), high-throughput scanning for DNA variations." *Hum Mol Genet 10*(16), 1657–1664.
25. Fan, J. B., Chen, X., Halushka, M. K., Berno, A., Huang, X., Ryder, T., Lipshutz, R. J., Lockhart, D. J., & Chakravarti, A. (2000). "Parallel genotyping of human SNPs using generic high-density oligonucleotide tag arrays." *Genome Res 10*(6), 853–860.
26. Fedurco, M., Romieu, A., Williams, S., Lawrence, I., & Turcatti, G. (2006). "BTA, a novel reagent for DNA attachment on glass and efficient generation of solid-phase amplified DNA colonies." *Nucleic Acids Res 34*(3), e22.
27. Ferguson, J. A., Steemers, F. J., & Walt, D. R. (2000). "High-density fiber-optic DNA random microsphere array." *Anal Chem 72*(22), 5618–5624.
28. Flusberg, B. A., Webster, D. R., Lee, J. H., Travers, K. J., Olivares, E. C., Clark, T. A., Korlach, J., & Turner, S. W. (2010). "Direct detection of DNA methylation during single-molecule, real-time sequencing." *Nat Methods 7*(6), 461–465.
29. Gaglio, D., Capitano, F., Mastrodonato, A., Minicocci, E., Deiana, C., Fragapane, P., Camilloni, G., & Mele, A. (2014). "Learning induced epigenetic modifications in the ventral striatum are necessary for long-term memory." *Behav Brain Res 265*, 61–68.
30. Gibbs, R. A., Nguyen, P. N., & Caskey, C. T. (1989). "Detection of single DNA base differences by competitive oligonucleotide priming." *Nucleic Acids Res* ***17***(7), 2437–2448.

31. Godovac-Zimmermann, J., & Brown, L. R. (2001). "Perspectives for mass spectrometry and functional proteomics." *Mass Spectrom Rev 20*(1), 1–57.
32. Goodwin, S., McPherson, J. D., & McCombie, W. R. Coming of age: ten years of next-generation sequencing technologies. Nat Rev Genet. 2016 May 17;17(6):333-51.
33. Gowda, M., Jantasuriyarat, C., Dean, R. A., & Wang, G. L. (2004). "Robust-LongSAGE (RL-SAGE): a substantially improved LongSAGE method for gene discovery and transcriptome analysis." *Plant Physiol 134*(3), 890–897.
34. Griffin, T. J., Tang, W., & Smith, L. M. (1997). "Genetic analysis by peptide nucleic acid affinity MALDI-TOF mass spectrometry." *Nat Biotechnol 15*(13), 1368–1372.
35. Gupta, S., Manubhai, K. P., Kulkarni, V., & Srivastava, S. (2016). An overview of innovations and industrial solutions in Protein Microarray Technology. *Proteomics 16*(8), 1297–1308.
36. Haff, L. A., & Smirnov, I. P. (1997). "Single-nucleotide polymorphism identification assays using a thermostable DNA polymerase and delayed extraction MALDI-TOF mass spectrometry." *Genome Res 7*(4), 378–388.
37. Hall, J. G., Eis, P. S., Law, S. M., Reynaldo, L. P., Prudent, J. R., Marshall, D. J., Allawi, H. T., Mast, A. L., Dahlberg, J. E., Kwiatkowski, R. W., de Arruda, M., Neri, B. P., & Lyamichev, V. I. (2000). "Sensitive detection of DNA polymorphisms by the serial invasive signal amplification reaction." *Proc Natl Acad Sci U S A 97*(15), 8272–8277.
38. Hardenbol, P., Banér, J., Jain, M., Nilsson, M., Namsaraev, E. A., Karlin-Neumann, G. A., Fakhrai-Rad, H., Ronaghi, M., Willis, T. D., Landegren, U., & Davis, R. W. (2003). "Multiplexed genotyping with sequence-tagged molecular inversion probes." *Nat Biotechnol 21*(6), 673–678.
39. Hardenbol, P., Yu, F., Belmont, J., Mackenzie, J., Bruckner, C., Brundage, T., Boudreau, A., Chow, S., Eberle, J., Erbilgin, A., Falkowski, M., Fitzgerald, R., Ghose, S., Iartchouk, O., Jain, M., Karlin-Neumann, G., Lu, X., Miao, X., Moore, B., Moorhead, M., Namsaraev, E., Pasternak, S., Prakash, E., Tran, K., Wang, Z., Jones, H. B., Davis, R. W., Willis, T. D., & Gibbs, R. A. (2005). "Highly multiplexed molecular inversion probe genotyping: over 10,000 targeted SNPs genotyped in a single tube assay." *Genome Res 15*(2), 269–275.

40. Herman, J. G., Graff, J. R., Myöhänen, S., Nelkin, B. D., & Baylin, S. B. (1996). "Methylation-specific PCR: a novel PCR assay for methylation status of CpG islands." *Proc Natl Acad Sci U S A 93*(18), 9821–9826.
41. Hirschhorn, J. N., Sklar, P., Lindblad-Toh, K., Lim, Y. M., Ruiz-Gutierrez, M., Bolk, S., Langhorst, B., Schaffner, S., Winchester, E., & Lander, E. S. (2000). "SBE-TAGS: an array-based method for efficient single-nucleotide polymorphism genotyping." *Proc Natl Acad Sci U S A 97*(22), 12164–12169.
42. Holland, P. M., Abramson, R. D., Watson, R., & Gelfand, D. H. (1991). "Detection of specific polymerase chain reaction product by utilizing the 5'----3' exonuclease activity of Thermus aquaticus DNA polymerase." *Proc Natl Acad Sci U S A 88*(16), 7276–7280.
43. Howell, W. M., Jobs, M., Gyllensten, U., & Brookes, A. J. (1999). "Dynamic allele-specific hybridization. A new method for scoring single nucleotide polymorphisms." *Nat Biotechnol 17*(1), 87–88.
44. Hu, M., Yao, J.,and Polyak, K. (2006). "Methylation-specific digital karyotyping." *Nat Protoc 1*(3), 1621–1636.
45. Jaligot, E., Beulé, T.,and Rival, A. (2002). "Methylation-sensitive RFLPs: characterisation of two oil palm markers showing somaclonal variation-associated polymorphism." *Theor Appl Genet 104*(8), 1263–1269.
46. Kang, K. A., Piao, M. J., Kim, K. C., Kang, H. K., Chang, W. Y., Park, I. C., Keum, Y. S., Surh, Y. J., & Hyun, J. W. (2014). "Epigenetic modification of Nrf2 in 5-fluorouracil-resistant colon cancer cells: involvement of TET-dependent DNA demethylation." *Cell Death Dis 5*, e1183.
47. Karas, M., & Hillenkamp, F. (1988). "Laser desorption ionization of proteins with molecular masses exceeding 10,000 daltons." *Anal Chem 60*(20), 2299–2301.
48. Karimi, M., Johansson, S., Stach, D., Corcoran, M., Grandér, D., Schalling, M., Bakalkin, G., Lyko, F., Larsson, C., & Ekström, T. J. (2006). "LUMA (LUminometric Methylation Assay)--a high throughput method to the analysis of genomic DNA methylation." *Exp Cell Res 312*(11), 1989–1995.
49. Kim, J. B., Porreca, G. J., Song, L., Greenway, S. C., Gorham, J. M., Church, G. M., Seidman, C. E., & Seidman, J. G. (2007). "Polony multiplex analysis of gene expression (PMAGE) in mouse hypertrophic

cardiomyopathy." *Science 316*(5830), 1481–1484.

50. Kim, S., & Misra, A. (2007). "SNP genotyping: technologies and biomedical applications." *Annu Rev Biomed Eng 9*, 289–320.

51. Kim, S., Edwards, J. R., Deng, L., Chung, W., & Ju, J. (2002). "Solid phase capturable dideoxynucleotides for multiplex genotyping using mass spectrometry." *Nucleic Acids Res 30*(16), e85.

52. Knierim, E., Lucke, B., Schwarz, J. M., Schuelke, M., & Seelow, D. (2011). "Systematic comparison of three methods for fragmentation of long-range PCR products for next generation sequencing." PLoS One **6**(11), e28240.

53. Kukita, Y., Higasa, K., Baba, S., Nakamura, M., Manago, S., Suzuki, A., Tahira, T., & Hayashi, K. (2002). "A single-strand conformation polymorphism method for the large-scale analysis of mutations/polymorphisms using capillary array electrophoresis." Electrophoresis **23**(14), 2259–2266.

54. Kuo, K. C., McCune, R. A., Gehrke, C. W., Midgett, R., & Ehrlich, M. (1980). "Quantitative reversed-phase high performance liquid chromatographic determination of major and modified deoxyribonucleosides in DNA." *Nucleic Acids Res 8*(20), 4763–4776.

55. Kurdyukov, S., and Bullock, M. (2016). "DNA Methylation Analysis: Choosing the Right Method." *Biology (Basel) 5*(1).

56. Landegren, U., Kaiser, R., Sanders, J., & Hood, L. (1988). "A ligase-mediated gene detection technique." *Science 241*(4869), 1077–1080.

57. Lander, E. S., Linton, L. M., Birren, B., Nusbaum, C., Zody, M. C., Baldwin, J., et al. (2001). "Initial sequencing and analysis of the human genome." *Nature 409*(6822), 860–921.

58. Lappalainen, T., Sammeth, M., Friedländer, M. R., 't Hoen, P. A., Monlong, J., Rivas, M. A., et al. (2013). "Transcriptome and genome sequencing uncovers functional variation in humans." *Nature 501*(7468), 506–511.

59. Laszlo, A. H., Derrington, I. M., Brinkerhoff, H., Langford, K. W., Nova, I. C., Samson, J. M., Bartlett, J. J., Pavlenok, M., & Gundlach, J. H. (2013). "Detection and mapping of 5-methylcytosine and 5-hydroxymethylcytosine with nanopore MspA." *Proc Natl Acad Sci U S A 110*(47), 18904–18909.

60. Latorra, D., Campbell, K., Wolter, A., & Hurley, J. M. (2003). "Enhanced allele-specific PCR discrimination in SNP genotyping using 3' locked

nucleic acid (LNA) primers." *Hum Mutat 22*(1), 79–85.

61. Le Hellard, S., Ballereau, S. J., Visscher, P. M., Torrance, H. S., Pinson, J., Morris, S. W., Thomson, M. L., Semple, C. A., Muir, W. J., Blackwood, D. H., Porteous, D. J., & Evans, K. L. (2002). "SNP genotyping on pooled DNAs: comparison of genotyping technologies and a semi automated method for data storage and analysis." *Nucleic Acids Res 30*(15), e74.
62. Leamon, J. H., Lee, W. L., Tartaro, K. R., Lanza, J. R., Sarkis, G. J., deWinter, A. D., Berka, J., Weiner, M., Rothberg, J. M., & Lohman, K. L. (2003). "A massively parallel PicoTiterPlate based platform for discrete picoliter-scale polymerase chain reactions." *Electrophoresis 24*(21), 3769–3777.
63. Lee, Y. K., Jin, S., Duan, S., Lim, Y. C., Ng, D. P., Lin, X. M., Yeo,G. S. H., & Ding, C. (2014). "Improved reduced representation bisulfite sequencing for epigenomic profiling of clinical samples." *Biol Proced Online 16*(1), 1.
64. Levene, M. J., Korlach, J., Turner, S. W., Foquet, M., Craighead, H. G., & Webb, W. W. (2003). "Zero-mode waveguides for single-molecule analysis at high concentrations." *Science 299*(5607), 682–686.
65. Levin, J. Z., Yassour, M., Adiconis, X., Nusbaum, C., Thompson, D. A., Friedman, N., Gnirke, A., & Regev, A. (2010). "Comprehensive comparative analysis of strand-specific RNA sequencing methods." *Nat Methods 7*(9), 709–715.
66. Li, W. H. & Sadler, L. A. (1991). "Low nucleotide diversity in man." *Genetics 129*(2), 513–523.
67. Liu, L., Li, Y., Li, S., Hu, N., He, Y., Pong, R., Lin, D., Lu, L., & Law, M. (2012). "Comparison of next-generation sequencing systems." *J Biomed Biotechnol 2012*, 251364.
68. Lizardi, P. M., Huang, X., Zhu, Z., Bray-Ward, P., Thomas, D. C., & Ward, D. C. (1998). "Mutation detection and single-molecule counting using isothermal rolling-circle amplification." *Nat Genet 19*(3), 225–232.
69. Lyamichev, V., Mast, A. L., Hall, J. G., Prudent, J. R., Kaiser, M. W., Takova, T., Kwiatkowski, R. W., Sander, T. J., M. de Arruda, Arco, D. A., Neri, B. P., & Brow, M. A. (1999). "Polymorphism identification and quantitative detection of genomic DNA by invasive cleavage of oligonucleotide probes." *Nat Biotechnol 17*(3), 292–296.

70. MacBeath, G. (2002). "Protein microarrays and proteomics." *Nat Genet 32 Suppl*, 526–532.

71. Mahapatra, S., Klee, E. W., Young, C. Y., Sun, Z., Jimenez, R. E., Klee, G. G., Tindall, D. J., & Donkena, K. V. (2012). "Global methylation profiling for risk prediction of prostate cancer." *Clin Cancer Res 18*(10), 2882–2895.

72. Margulies, M., Egholm, M., Altman, W. E., Attiya, S., Bader, J. S., Bemben, L. A., et al. (2005). "Genome sequencing in microfabricated high-density picolitre reactors." *Nature 437*(7057), 376–380.

73. Mass Spectrometry for Proteomics. *Neuroproteomics*. O. Alzate. Boca Raton (FL), CRC CRC Press/Taylor & Francis; 2010. Chapter 5.

74. Matsumura, H., Reich, S., Ito, A., Saitoh, H., Kamoun, S., Winter, P., Kahl, G., Reuter, M., Kruger, D. H., & Terauchi, R. (2003). "Gene expression analysis of plant host-pathogen interactions by SuperSAGE." *Proc Natl Acad Sci U S A 100*(26), 15718–15723.

75. Matsuzaki, H., Dong, S., Loi, H., Di, X., Liu, G., Hubbell, E., Law, J., Berntsen, T., Chadha, M., Hui, H., Yang, G., Kennedy, G. C., Webster, T. A., Cawley, S., Walsh, P. S., Jones, K. W., Fodor, S. P., & Mei, R. (2004). "Genotyping over 100,000 SNPs on a pair of oligonucleotide arrays." *Nat Methods 1*(2), 109–111.

76. Matsuzaki, H., Loi, H., Dong, S., Tsai, Y. Y., Fang, J., Law, J., Di, X., Liu, W. M., Yang, G., Liu, G., Huang, J., Kennedy, G. C., Ryder, T. B., Marcus, G. A., Walsh, P. S., Shriver, M. D., Puck, J. M., Jones, K. W., & Mei, R. (2004). "Parallel genotyping of over 10,000 SNPs using a one-primer assay on a high-density oligonucleotide array." *Genome Res 14*(3), 414–425.

77. Maxam, A. M., & Gilbert, W. (1977). "A new method for sequencing DNA." *Proc Natl Acad Sci U S A 74*(2), 560–564.

78. McClay, J. L., Sugden, K., Koch, H. G., Higuchi, S., & Craig, I. W. (2002). "High-throughput single-nucleotide polymorphism genotyping by fluorescent competitive allele-specific polymerase chain reaction (SNiPTag)." *Anal Biochem 301*(2), 200–206.

79. Medintz, I., Wong, W. W., Berti, L., Shiow, L., Tom, J., Scherer, J., Sensabaugh, G., & Mathies, R. A. (2001). "High-performance multiplex SNP analysis of three hemochromatosis-related mutations with capillary array electrophoresis microplates." *Genome Res 11*(3), 413–421.

80. Meissner, A., Gnirke, A., Bell, G. W., Ramsahoye, B., Lander, E. S., & Jaenisch, R. (2005). "Reduced representation bisulfite sequencing for comparative high-resolution DNA methylation analysis." *Nucleic Acids Res 33*(18), 5868–5877.

81. Metzker, M. L. (2010). "Sequencing technologies – the next generation." *Nat Rev Genet 11*(1), 31–46.

82. Miller, M. B., & Tang, Y. W. (2009). "Basic concepts of microarrays and potential applications in clinical microbiology." *Clin Microbiol Rev 22*(4), 611–633.

83. Nickerson, D. A., Taylor, S. L., Weiss, K. M., Clark, A. G., Hutchinson, R. G., Stengård, J., Salomaa, V., Vartiainen, E., Boerwinkle, E., & Sing, C. F. (1998). "DNA sequence diversity in a 9.7-kb region of the human lipoprotein lipase gene." *Nat Genet 19*(3), 233–240.

84. Nikiforov, T. T., Rendle, R. B., Goelet, P., Rogers, Y. H., Kotewicz, M. L., Anderson, S., Trainor, G. L., & Knapp, M. R.(1994). "Genetic Bit Analysis: a solid phase method for typing single nucleotide polymorphisms." *Nucleic Acids Res 22*(20), 4167–4175.

85. Nilsson, M., Malmgren, H., Samiotaki, M., Kwiatkowski, M., Chowdhary, B. P., & Landegren, U. (1994). "Padlock probes: circularizing oligonucleotides for localized DNA detection." *Science 265*(5181), 2085–2088.

86. Ogino, S., Nosho, K., Kirkner, G. J., Kawasaki, T., Chan, A. T., Schernhammer, E. S., Giovannucci, E. L., & Fuchs, C. S. (2008). "A cohort study of tumoral LINE-1 hypomethylation and prognosis in colon cancer." *J Natl Cancer Inst 100*(23), 1734–1738.

87. Orita, M., Suzuki, Y., Sekiya, T., & Hayashi, K. (1989). "Rapid and sensitive detection of point mutations and DNA polymorphisms using the polymerase chain reaction." *Genomics 5*(4), 874–879.

88. Pan, Q., Shai, O., Lee, L. J., Frey, B. J., & Blencowe, B. J. (2008). "Deep surveying of alternative splicing complexity in the human transcriptome by high-throughput sequencing." *Nat Genet 40*(12), 1413–1415.

89. Parker, C. E., Warren, M. R., & Mocanu, V. Francis; 2010. Chapter 6. Frontiers in Neuroscience.

90. Pastinen, T., Kurg, A., Metspalu, A., Peltonen, L., & Syvänen, A. C. (1997). "Minisequencing: a specific tool for DNA analysis and diagnostics on oligonucleotide arrays." *Genome Res 7(*6), 606–614.

91. Phizicky, E., Bastiaens, P. I., Zhu, H., Snyder, M., & Fields, S. (2003). "Protein analysis on a proteomic scale." *Nature 422*(6928), 208–215.

92. Piétu, G., Mariage-Samson, R., Fayein, N. A., Matingou, C., Eveno, E., Houlgatte, R., Decraene, C., Vandenbrouck, Y., Tahi, F., Devignes, M. D., Wirkner, U., Ansorge, W., Cox, D., Nagase, T., Nomura, N., & Auffray, C. (1999). "The Genexpress IMAGE knowledge base of the human brain transcriptome: a prototype integrated resource for functional and computational genomics." *Genome Res 9*(2), 195–209.

93. Quail, M. A., Smith, M., Coupland, P., Otto, T. D., Harris, S. R., Connor, T. R., Bertoni, A., Swerdlow, H. P., & Gu, Y. (2012). "A tale of three next generation sequencing platforms: comparison of Ion Torrent, Pacific Biosciences and Illumina MiSeq sequencers." *BMC Genomics 13*, 341.

94. Ramachandran, N., Hainsworth, E., Demirkan, G.,and LaBaer, J. (2006). "On-chip protein synthesis for making microarrays." *Methods Mol Biol 328*, 1–14.

95. Rocha, D., Gut, I., Jeffreys, A. J., Kwok, P. Y., Brookes, A. J., & Chanock, S. J. (2006). "Seventh international meeting on single nucleotide polymorphism and complex genome analysis: 'ever bigger scans and an increasingly variable genome.'" *Hum Genet 119*(4), 451–456.

96. Ronaghi, M., Karamohamed, S., Pettersson, B., Uhlén, M., & Nyrén, P. (1996). "Real-time DNA sequencing using detection of pyrophosphate release." *Anal Biochem 242*(1), 84–89.

97. Ross, P., Hall, L., Smirnov, I., & Haff, L. (1998). "High level multiplex genotyping by MALDI-TOF mass spectrometry." *Nat Biotechnol 16*(13), 1347–1351.

98. Rothberg, J. M., Hinz, W., Rearick, T. M., Schultz, J., Mileski, W., Davey, M., et al. (2011). "An integrated semiconductor device enabling non-optical genome sequencing." *Nature 475*(7356), 348–352.

99. Russom, A., Haasl, S., Brookes, A. J., Andersson, H., & Stemme, G. (2006). "Rapid melting curve analysis on monolayered beads for high-throughput genotyping of single-nucleotide polymorphisms." *Anal Chem 78*(7), 2220–2225.

100. Sachidanandam, R., Weissman, D., Schmidt, S. C., Kakol, J. M., Stein, L. D., Marth, G., et al. (2001). "A map of human genome sequence variation containing 1.42 million single nucleotide polymorphisms."

Nature 409(6822), 928–933.

101. Saha, S., Sparks, A. B., Rago, C., Akmaev, V., Wang, C. J., Vogelstein, B., Kinzler, K. W., & Velculescu, V. E. (2002). "Using the transcriptome to annotate the genome." *Nat Biotechnol 20*(5), 508–512.
102. Sanchez, J. J., Børsting, C., Hallenberg, C., Buchard, A., Hernandez, A., & Morling, N. (2003). "Multiplex PCR and minisequencing of SNPs--a model with 35 Y chromosome SNPs." *Forensic Sci Int 137*(1), 74–84.
103. Sanger, F., Nicklen, S., & Coulson, A. R. (1977). "DNA sequencing with chain-terminating inhibitors." P*roc Natl Acad Sci U S A 74*(12), 5463–5467.
104. Sasakura, Y., Kanda, K., Yoshimura-Suzuki, T., Matsui, T., Fukuzono, S., Han, M. H., & Shimizu, T. (2004). "Protein microarray system for detecting protein-protein interactions using an anti-His-tag antibody and fluorescence scanning: effects of the heme redox state on protein-protein interactions of heme-regulated phosphodiesterase from Escherichia coli." *Anal Chem 76*(22), 6521–6527.
105. Sauer, S., Gelfand, D. H., Boussicault, F., Bauer, K., Reichert, F., & Gut, I. G. (2002). "Facile method for automated genotyping of single nucleotide polymorphisms by mass spectrometry." *Nucleic Acids Res 30*(5), e22.
106. Schreiber, J., Wescoe, Z. L., Abu-Shumays, R., Vivian, J. T., Baatar, B., Karplus, K., & Akeson, M. (2013). "Error rates for nanopore discrimination among cytosine, methylcytosine, and hydroxymethylcytosine along individual DNA strands." *Proc Natl Acad Sci U S A 110*(47), 18910–18915.
107. Shapero, M. H., Leuther, K. K., Nguyen, A., Scott, M., & Jones, K. W. (2001). "SNP genotyping by multiplexed solid-phase amplification and fluorescent minisequencing." *Genome Res 11*(11), 1926–1934.
108. Shen, R., Fan, J. B., Campbell, D., Chang, W., Chen, J., Doucet, D., Yeakley, J., Bibikova, M., Wickham, E.,Garcia, McBride, C., Steemers, F., Garcia, F., Kermani, B. G., Gunderson, K., & Oliphant, A. (2005). "High-throughput SNP genotyping on universal bead arrays." *Mutat Res 573*(1–2), 70–82.
109. Shiu, S. H., & Borevitz, J. O.(2008). "The next generation of microarray research: applications in evolutionary and ecological genomics." *Heredity (Edinb) 100*(2), 141–149.

110. Silzel, J. W., Cercek, B., Dodson, C., Tsay, T., & Obremski, R. J. (1998). "Mass-sensing, multianalyte microarray immunoassay with imaging detection." *Clin Chem 44*(9), 2036–2043.

111. Sokolov, B. P. (1990). "Primer extension technique for the detection of single nucleotide in genomic DNA." *Nucleic Acids Res 18*(12), 3671.

112. Song, L., James, S. R., Kazim, L., & Karpf, A. R. (2005). "Specific method for the determination of genomic DNA methylation by liquid chromatography-electrospray ionization tandem mass spectrometry." *Anal Chem 77*(2), 504–510.

113. Sultan, M., Schulz, M. H., Richard, H., Magen, A., Klingenhoff, A., Scherf, M., Seifert, M., Borodina, T., Soldatov, A., Parkhomchuk, D., Schmidt, D., O'Keeffe, S., Haas, S., Vingron, M., Lehrach, H., & Yaspo, M. L. (2008). "A global view of gene activity and alternative splicing by deep sequencing of the human transcriptome." *Science* ***321***(5891), 956–960.

114. Syed, P., Gupta, S., Choudhary, S., Pandala, N. G., Atak, A., Richharia, A., M. K P, Zhu, H., Epari, S., Noronha, S. B., Moiyadi, A., & Srivastava, S. (2015). "Autoantibody Profiling of Glioma Serum Samples to Identify Biomarkers Using Human Proteome Arrays." *Sci Rep 5*, 13895.

115. Talapatra, A., Rouse, R., & Hardiman, G. (2002). "Protein microarrays: challenges and promises." *Pharmacogenomics 3*(4), 527–536.

116. Tarasova, G. V., Nayakshina, T. N., & Degtyarev, S. K. (2008). "Substrate specificity of new methyl-directed DNA endonuclease GlaI." *BMC Mol Biol 9*, 7.

117. Tobler, A. R., Short, S., Andersen, M. R., Paner, T. M., Briggs, J. C., Lambert, S. M., Wu, P. P., Wang, Y., Spoonde, A. Y., Koehler, R. T., Peyret, N., Chen, C., Broomer, A. J., Ridzon, D. A., Zhou, H., Hoo, B. S., Hayashibara, K. C., Leong, L. N., Ma, C. N., Rosenblum, B. B., Day, J. P., Ziegle, J. S., F. M. De La Vega, Rhodes, M. D., Hennessy, K. M., & Wenz, H. M. (2005). "The SNPlex genotyping system: a flexible and scalable platform for SNP genotyping." *J Biomol Tech 16*(4), 398–406.

118. Tong, A. K., Li, Z., Jones, G. S., Russo, J. J., & Ju, J. (2001). "Combinatorial fluorescence energy transfer tags for multiplex biological assays." *Nat Biotechnol 19*(8), 756–759.

119. Ugozzoli, L., Wahlqvist, J. M., Ehsani, A., Kaplan, B. E., & Wallace, R. B. (1992). "Detection of specific alleles by using allele-specific primer extension followed by capture on solid support." *Genet Anal Tech Appl 9*(4), 107–112.

120. Velculescu, V. E., Zhang, L., Vogelstein, B., & Kinzler, K. W. (1995). "Serial analysis of gene expression." *Science 270*(5235), 484–487.

121. Velculescu, V. E., Zhang, L., Zhou, W., Vogelstein, J., Basrai, M. A., Bassett, D. E., Hieter, P., Vogelstein, B., & Kinzler, K. W. (1997). "Characterization of the yeast transcriptome." *Cell 88*(2), 243–251.

122. Venter, J. C., Adams, M. D., Myers, E. W., Li, P. W., Mural, R. J., Sutton, G. G., et al. (2001). "The sequence of the human genome." *Science 291*(5507), 1304–1351.

123. Villaseñor-Park, J., & Ortega-Loayza, A. G. (2013). "Microarray technique, analysis, and applications in dermatology." *J Invest Dermatol 133*(4), e7.

124. Walker, D. L., Bhagwate, A. V., Baheti, S., Smalley, R. L., Hilker, C. A., Sun, Z., & Cunningham, J. M. (2015). "DNA methylation profiling: comparison of genome-wide sequencing methods and the Infinium Human Methylation 450 Bead Chip." *Epigenomics 7*(8), 1287–1302.

125. Wang, D. G., Fan, J. B., Siao, C. J., Berno, A., Young, P., Sapolsky, R., et al. (1998). "Large-scale identification, mapping, and genotyping of single-nucleotide polymorphisms in the human genome." *Science 280*(5366), 1077–1082.

126. Wang, Z., Gerstein, M., & Snyder, M. (2009). "RNA-Seq: a revolutionary tool for transcriptomics." *Nat Rev Genet 10*(1), 57–63.

127. Wilkins, M. R., Gasteiger, E., Gooley, A. A., Herbert, B. R., Molloy, M. P., Binz, P. A., Ou, K., Sanchez, J. C., Bairoch, A., Williams, K. L. & Hochstrasser, D. F. (1999). "High-throughput mass spectrometric discovery of protein post-translational modifications." *J Mol Biol 289*(3), 645–657.

128. Wilkins, M. R., Sanchez, J. C., Gooley, A. A., Appel, R. D., Humphery-Smith, I., Hochstrasser, D. F., & Williams, K. L. (1996). "Progress with proteome projects: why all proteins expressed by a genome should be identified and how to do it." *Biotechnol Genet Eng Rev 13*, 19–50.

129. Wojdacz, T. K., Dobrovic, A., & Hansen, L. L. (2008). "Methylation-sensitive high-resolution melting." *Nat Protoc 3*(12), 1903–1908.

130. Xu, H., Sha, M. Y., Wong, E. Y., Uphoff, J., Xu, Y., Treadway, J. A., Truong, A., O'Brien, E., Asquith, S., Stubbins, M., Spurr, N. K., Lai, E. H., & Mahoney, W. (2003). "Multiplexed SNP genotyping using the Qbead system: a quantum dot-encoded microsphere-based assay." *Nucleic Acids Res 31*(8), e43.

131. Yamamoto, M., Wakatsuki, T., Hada, A., & Ryo, A. (2001). "Use of serial analysis of gene expression (SAGE) technology." *J Immunol Methods 250*(1–2), 45–66.

CHAPTER 3

PROTEOMICS BIOINFORMATICS TOOLS IN PROTEOGENOMICS STUDIES

CONTENTS

3.1. INTRODUCTION

The major aim of proteomics is to characterize and evaluate the whole set of proteins existing in a sample at a certain time point. In order to reach this target, the most recommended approach is to digest protein into peptides followed by the careful analysis by liquid chromatography which is further coupled to tandem mass spectrometry (MS) (Aebersold and Mann, 2003). Such classical approach is always followed by a massive bioinformatic analysis which is employed to infer the sequences of different and ultimately posttranslational modifications (PTMs). From these data a list of proteins existing in the sample with their respective abundances can be inferred (Altelaar et al., 2013).

Specifically, there are three basic study designs are employed for characterization of peptide sequences from tandem mass spectrometry (Vaudel et al., 2012b):

- Spectral library searching: here experimental spectra are matched to a library of already recognized spectra, and the degree of the matches is assessed to allow for the inference of the existence of known peptides.
- *De novo* sequencing: here a sequence or part of a sequence is inferred from the spectrum by matching the distance between peaks to PTM masses and possible amino acids.
- Sequence databases searching: In this approach, theoretical spectrums are extrapolated from a database of expected protein sequences, and the match between experimental and theoretical spectra is in turn assessed to infer the existence of peptides.

Significantly, in most cases, mapping the peptide sequence back to the protein level through a reference protein sequence database is used, a method called protein inference (Nesvizhskii and Aebersold, 2005). However, as such spectral library mining can only yield peptides which have been formerly measured; it will be employed mainly for targeted protein quantification, not for discovery studies (Domon and Aebersold, 2006). In relation, the constant advances in instrumentation, the information offered in the spectrum is generally not sufficient for *de novo* sequencing to compete with searching the sequence databases. Therefore, sequence database mining tools, or the so-called search engines, are by far the most employed approach for the identification of peptides.

3.2. DATABASE SEARCH ENGINES

Proteomic search engines receive its input as:

(i) Peak lists from mass spectrometry experiments.

The handling of raw mass spectrometry data before further analysis classically comprises signal processing to denoise spectra, declining the baseline, and peak picking performing, in other words, transforming the profile spectra or the linear spectra of bell-shaped peaks, into the centroided spectra or the discrete peak lists (Lange et al., 2006; Vaudel et al., 2010). Yet, with the major advances of high resolution mass spectrometers comprising built-in signal processing units, this way has been hugely simplified, and without additional processing, such raw data can often simply be used.

As a part of the ProteoWizard package, this conversion of raw mass spectrometry files to open formats can be easily conducted by msconvert (Kessner et al., 2008) In addition, signal processing handling methods such as peak picking, if needed, can be applied within the OpenMS package (Kohlbacher et al., 2007) or by ProteoWizard (French et al., 2015).

(ii) Expected protein sequences.

Numerous open and public resources with established protein sequences are accessible. Some of these resources are specialized in terms of disease, sub-proteome or whole organism, while others are more general trying to provide complete collections of all recognized sequences. The foremost general protein databases are the UniProt knowledgebase (Apweiler et al., 2004), the closely related Ensembl database (Hubbard et al., 2002; Yates et al., 2016), the National Center for Biotechnology Information (NCBI) Protein database (Coordinators, 2016) delivering sequences from multiple sources comprising the Reference Sequence (RefSeq) Database (Pruitt et al., 2005)), and finally the DNA Data Bank of Japan (Tateno et al., 2002).

(iii) User defined search parameters.

Concluded from the search parameters, the user plans the search space of the algorithm. In other words, using the fragment ions and peptides which are expected from the given theoretical protein sequences, and controls the degrees of freedom with which the algorithm couples experimental mass spectrum to these theoretical ions and peptides. The parameters classically comprise: (i) information on how would be cleaved with a maximum limit on the count of missed cleavages, (ii) the predictable fragmentation behavior of ions and peptides, and (iii) predicted PTMs. Based on their prevalence, PTMs can be steadily accounted for, or the known as *fixed* or *static* modifications, or the algorithm can restate the possible modification statuses of a peptide, which is called the dynamic or the *variable* modifications.

Putting all these settings together, the algorithm will compare each experimental spectrum to the expected fragmentation pattern of each possible peptide, and for every spectrum, a back yield of a list called peptide spectrum matches (PSMs) along with a score given for each match. As the search space goes larger, this task becomes more complex. Therefore, one of the main challenges when using search engines in proteogenomics studies is the management of the search space.

Over the past decades, various algorithms were established, most of them are enumerated at the 'OMIC tools' web site (Henry et al., 2014) (omictools.com/database-search-category). Some of them have been encompassed into generic proteomic software suites like OpenMS (Kohlbacher et al., 2007), the Trans Proteomics Pipeline (TPP) (Deutsch et al., 2010) and MaxQuant (Cox and Mann, 2008). Otherwise, a more simple approach to operate search engines is to use them *via* SearchGUI (Vaudel et al., 2011a), Comet (Eng et al., 2013), X!Tandem (Craig and Beavis, 2004), Andromeda (Cox et al., 2011), MyriMatch (Tabb et al., 2007), OMSSA (Craig et al., 2004), MS Amanda (Dorfer et al., 2014), MS-GF+ (Kim and Pevzner, 2014), and Tide (Diament and Noble, 2011) – Figure 3.1.

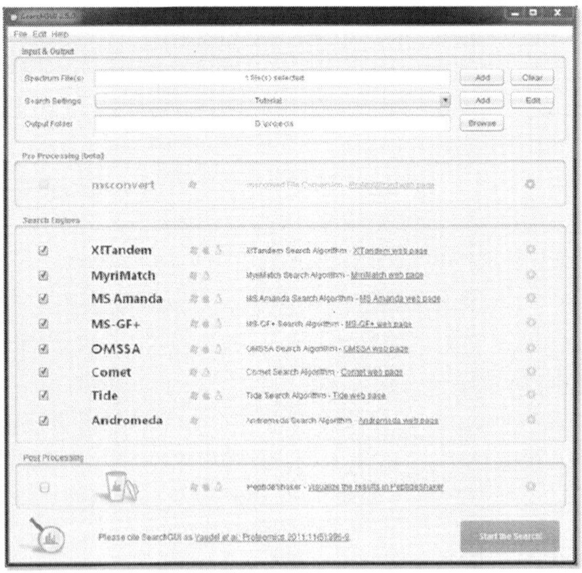

Figure 3.1. The key frame of SearchGUI permits the harnessing of various search engines.

3.3. RESULTS INTEGRATION

Result files are generated by each search engine, containing lists of PSMs, upon completion of the search. These lists are then assembled into proteins employing the post processing software tools. Among the critical tasks of such tools is to select a representative hit per spectrum and to use peptides to infer proteins (Nesvizhskii and Aebersold, 2005), and finally to estimate the rate of errors. Two methods exist for the evaluation of error rates, the so-called target-decoy strategy (Elias and Gygi, 2010) or the modeling of score distributions (Keller et al., 2002). In the target-decoy strategy, artificial decoy sequences are added to the sequence database or searched in parallel.

To model the distribution of the scores of false positive hits to derive the rate of errors, the distribution of the decoy scores is then used. These can be estimated for the entire dataset, providing a global false discovery rate (FDR) or locally, offering a probability for an identified compound to be a false result, a posterior error probability (PEP). It is also possible to do manual or semi-manual inspection of the results (Helsens et al., 2008) in addition to the statistical evaluation of error rates.

A simple way to aggregate the results of multiple search engines, apart from the software frameworks mentioned above, is to process them with a user friendly tool designed to work seamlessly in combination with SearchGUI, or the so called PeptideShaker (Vaudel et al., 2015). In addition to protein inference or peptide to protein mapping, error rates, and matches quality control, PeptideShaker tool performs PTM localization and quality filtering, all through interactive graphical user interface given, Figure 3.2.

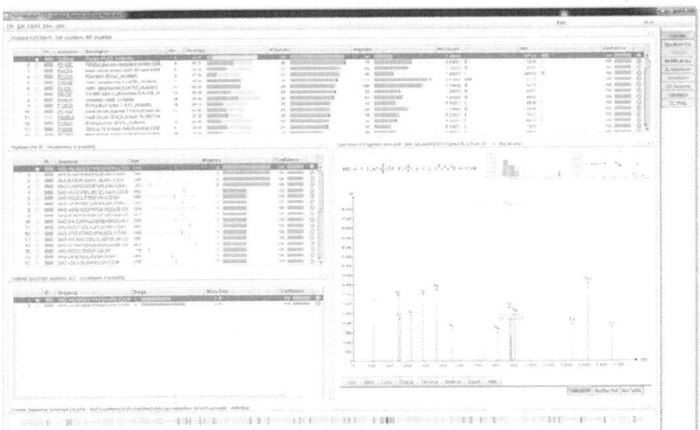

Figure 3.2. The graphical interface of PeptideShaker, as demonstrated here with the Overview tab, giving a global view of the recognized compounds.

Upon finishing the step of raw data processing, proteomics results interpretation intensely depends on their functional and biological context (Vaudel et al., 2014a). For example, data canbe mapped to public resources offering additional information (Vizcaino et al., 2009). In addition, it is also possible to map proteomic results to genetic information and therefore receiving knowledge at the genetic level or executing analysis of the gene ontology (GO) This can be done automatically in PeptideShaker when working with UniProt sequences. It is also possible to map peptide sequences and their modifications to the Protein Data Bank (PDB) (Sussman et al., 1998), and therefore yielding structural information at the level of the post-translational step.

Reactome, which allows for mapping to functional resources (Croft et al., 2011) enables further functional analyses. Additionally, proteomics data can be used in combination with the results and data derived from other modern omics technologies such as genomics, transcriptomics, and metabolomics. It is also mandatory, prior to publication, to share all data including the raw data files (Martens et al., 2005b; Barsnes and Martens, 2013). Therefore, repositories of proteomic data were initiated and coordinated through the environment of ProteomeXchange (Vizcaino et al., 2014; Martens et al., 2005a; Craig et al., 2004; Desiere et al., 2006).

3.4. LIMITATIONS IN DATABASE SEARCHING

Despite providing a unique throughput in the area of protein identification, searching the database searching also has several limits. The most important limitation is that the space of the search is limited to the content of the database being employed. In order to overcome such limitation and to increase the sample coverage by the database, it is probable to perform so-called error tolerant searches or to extend it to all expected proteins, including all isoforms, unanticipated cleavage, and all types of PTMs.

However, a big drawback of increasing the search space is the increased prevalence of false positives, in addition to the long search time. The false positives can be either random matches in the search space or partial matches. In partial matches, the error can be due to a minor detail of the peptide sequence, like the artefactual identification of a PTM. As mentioned, the presence of both classes of false positives correlates with the database size. Yet, while the presence of random matches can be sharply tracked through the target-decoy strategy (Vaudel et al., 2012a), it is a bit harder to assess the rate of non-random false positive matches (Colaert et al., 2011).

Furthermore, using very large databases to search decreases remarkably the distinguishability of the best scoring sequences, and the distinctiveness of protein mapping. Therefore, the post processing of these results will thus be more error prone and more complex. Hence, for better results, proteomic studies typically; limit the search space to the considered species; limit the inclusion of isoforms, of possible cleavage sites. In addition, processing the entire six-frame translation of a complex genome is therefore excluded.

On the other side of the scale, it can be recommended to modify the search space for a list of proteins of interest in order to reduce the processing time and avoid random error-prone matches. This paradoxically leads to forcing the algorithm to match specific sequences, and in the nonappearance of opposition between the targeted proteins and other possible nominees, the spectrum of peptides omitted from the database and resembling target sequences will result in the creation of non-random false matches.

As previously mentioned, these matches are not easily checked by error rates estimation classical methods (Colaert et al., 2011), and the final results will, therefore, be biased. This bias is remarkably the cause of the systematic inclusion of contaminant proteins sequences. The study on the bee colony collapse disorder represents a descriptive example of the problem that has received worldwide attention (Bromenshenk et al., 2010), where bee samples were searched with a potential pathogens database, resulting in the identification of pathogen from spectra more likely to be coming from bee proteins itself rather than the reporter fungi and viruses (Knudsen and Chalkley, 2011).

Thus the database search quality represents the delicate balance between covering all proteins present in the sample, and leave a quite small room for false positives. Therefore, low abundant species such as a variant of established proteins may fail to pass through the validation threshold or it can be easily missed by the identification procedure.

Searching with relaxed search settings before filtering represents one of the most prevalent approaches to ease this problem. Consequently, false positive results will be diverted outside the scope of the study and, furthermore using target and decoy hits distributions as a director to categorize areas of the search space with a high rate of true positives, it can be filtered out (Beausoleil et al., 2006; Vaudel et al., 2011b). This method will allow for the yield of more peptides especially highly efficient search strategies were employed to identify novel protein variants and if the decline of the false discovery rate pays off for the loss of hits due to competition in the augmented search space (Chick et al., 2015).

Using multiple pass searches, also named iterative searches, is an alternative way to overcome the problem of large search spaces. In this way, a smaller database of confidently identified proteins is constructed from the result of a first search. Then search again with more tolerance approach towards sequence variations and post-translational modifications, which is an approach offered by the X!Tandem search engine (Craig and Beavis, 2004). Targeted databases can also be constructed from public repositories to evade the challenging preliminary search, (Shanmugam and Nesvizhskii, 2015).

As a substitution, the result of the first search can be used to abandon spectra of established proteins. Then, the unidentified spectra are then searched (Noble, 2015). This method allows for the sequential searching of the databases of increasing complexity. However, the liability of introduction of non-random false hits, and the difficulty of applying the techniques of standard error rate estimation make the use of these search strategies debatable and needs further refinement (Everett et al., 2010; Bern and Kil, 2011).

3.5. APPLICATION TO PROTEOGENOMICS

In the hard trial to establish a balance between search space size and sample coverage, proteogenomics, from the study of the protein expression process, offers the unique solution to add additional relevant sequences. On the other hand, the identified sequences allow refining the annotation at both the transcriptomics and the genomic level (Jaffe et al., 2004). Therefore, Proteogenomics connects all fields of modern molecular biology involved in the study of protein expression, genomics, transcriptomics, and proteomics.

Nowadays, it has become more applicable to execute multiple omic analyses on the same set of samples through the integration of the use of next-generation sequencing (NGS) techniques and of high throughput mass spectrometry. Accordingly, ribosomal profiling, genome sequencing, and RNA sequencing can be employed to hypothesize the existence of specific variants like specific proteoforms. While reciprocally, protein identification methods can be used to settle their existence.

Such multiomics approaches allow for new spaces for more applications. For example, the exploration of samples comprising multiple or no defined species and the revision of organisms with little or short in the genetic annotation (Muth et al., 2013a). In human, a favorable application is the use of proteogenomics approaches to examine the expression of disease specific variants (Alfaro et al., 2014; Zhang et al., 2014), as detected in chemoresistant and chemoresponsive cancer cell lines (Pemovska et al., 2013), and inferring possible effects on the phenotype from there.

In order to put such multi omics approaches in use, the demand of novel bioinformatic tools was under strong demand. Therefore, a myriad of tools and resources were recently established to interpret the hunted information. These comprise the generation of databases of variants readily usable on proteomics datasets and non-categorized sequences (Sheynkman et al., 2014; Crappe et al., 2015; Risk et al., 2013), as reviewed in details by Nesvizhskii (2014). As a result, the entire proteogenomics description of samples is currently handled by standard available and organized workflows (Boekel et al., 2015).

Meanwhile, it should be stressed here that despite the high benefit of such bioinformatic tools, it requires a high level of bioinformatic expertise due to its high complexity. In addition, such bioinformatic tools and workflows will not run the standard desktop computers and would be preferred to run these programs as distributed tasks. (Afgan et al., 2012; Trudgian and Mirzaei, 2012; Muth et al., 2013b; Verheggen et al., 2014). Taking advantage of the powerful environment of GALAXY will allow for a better use of the proteogenomics workflow (Boekel et al., 2015; Fan et al., 2015; Jagtap et al., 2014; Sheynkman et al., 2014).

On of the major limitation of the proteogenomics approaches is the low abundance of the non-canonical gene products. Such low abundant species in addition to the lack of amplification method for amino acid sequences resulting in a lower possibility to trigger events in the mass spectrometry scan. Consequently, the potential of the discovery of novel products can be limited by the proteomic workflow sensitivity. It is thus mandatory to check very large amounts of data, to augment the possibilities of mining novel gene products (Whiteaker et al., 2014). This can be excuted by reprocessing publicly available datasets (Vaudel et al., 2016). sORFs.org (Olexiouk et al., 2016) and LNCipedia (Volders et al., 2015) are models of such approaches, in which the abovementioned tools are combined with automated search parameter inference (Hulstaert et al., 2013) and then applied to mine ProteomeXchange datasets in proteogenomics contexts (Verheggen et al., 2015). It is, however, worth noting that when using these kinds of big data strategies, the mentioned bioinformatic challenges such as global and local error rates estimation are augmented.

3.6. CONCLUSION

In conclusion, among the limiting factors of proteogenomics studies appear the limitations of database search engines and the reliable estimation of error rates as major players. A possible tip could be declining the dependence on the database, *through* better spectral library searching and integration of de novo sequencing. Spectral libraries can, by ruling out commonly observed peptides, effectively be employed to decrease the number of spectra searched. In big data approaches, where a large number of experiments are investigated for low abundant compounds, it is likely to cluster spectra preceding to processing, and mine these clusters of commonly met non identified spectra for novel gene products (Griss et al., 2013).

The evolution of proteogenomics opened a new slot for inspiring discoveries, and the distribution of these approaches will considerably augment the precision of characterization of biomedical samples. Presently, proteogenomics studies, however, necessitate specialized scientific know-how, particularly in bioinformatics, hindering the possible applicability of the approach. Efforts being currently devoted towards the development of user friendly and interactive workflows will definitely allow for the overcome of this restraint.

REFERENCES

1. Aebersold, R., & Mann, M. (2003). Mass spectrometry-based proteomics. *Nature, 422* (6928), 198–207.
2. Afgan, E., Chapman, B., & Taylor, J. (2012). CloudMan as a platform for tool, data, and analysis distribution. *BMC Bioinformatics, 13*, 315.
3. Alfaro, J. A., Sinha, A., Kislinger, T., & Boutros, P. C. (2014). Onco-proteogenomics: cancer proteomics joins forces with genomics. *Nature Methods, 11* (11), 1107–1113. Altelaar, A. F., Munoz, J., & Heck, A. J. (2013). Next generation proteomics: towards an integrative view of proteome dynamics. *Nature Reviews Genetics, 14* (1), 35–48.
4. Apweiler, R., Bairoch, A., Wu, C. H., Barker, W. C., Boeckmann, B., Ferro, S., Gasteiger, E., Huang, H., Lopez, R., Magrane, M., Martin, M. J., Natale, D. A., O'Donovan, C., Redaschi, N., & Yeh, L. S. (2004). UniProt: The Universal Protein knowledgebase. *Nucleic Acids Research, 32* (Database issue), D115–D119.
5. Barsnes, H., & Martens, L. (2013). Crowd sourcing in proteomics: Public resources lead to better experiments. *Amino Acids, 44* (4), 1129–1137.
6. Beausoleil, S. A., Villen, J., Gerber, S. A., Rush, J., & Gygi, S. P. (2006). A probability-based approach for high-throughput protein phosphorylation analysis and site localization. *Nature Biotechnology, 24* (10), 1285–1292.
7. Bern, M., & Kil, Y. J. (2011). Comment on "Unbiased statistical analysis for multi-stage proteomic search strategies." *Journal of Proteome Research, 10* (4), 2123–2127.
8. Boekel, J., Chilton, J. M., Cooke, I. R., Horvatovich, P. L., Jagtap, P. D., Kall, L., Lehtio, J., Lukasse, P., Moerland, P. D., & Griffin, T. J. (2015). Multi-omic data analysis using Galaxy. *Nature Biotechnology, 33* (2), 137–139.
9. Bromenshenk, J. J., Henderson, C. B., Wick, C. H., Stanford, M. F., Zulich, A. W., Jabbour, R. E., Deshpande, S. V., McCubbin, P. E., Seccomb, R. A., Welch, P. M., Williams, T., Firth, D. R., Skowronski, E., Lehmann, M. M., Bilimoria, S. L., Gress, J., Wanner, K. W., & Cramer, R. A., Jr. (2010). Iridovirus and microsporidian linked to honey bee colony decline. *PLoS One, 5* (10), e13181.
10. Chick, J. M., Kolippakkam, D., Nusinow, D. P., Zhai, B., Rad, R., Huttlin, E. L., & Gygi, S. P. (2015). A mass tolerant database search

identifies a large proportion of unassigned spectra in shotgun proteomics as modified peptides. *Nature Biotechnology, 33* (7), 743–749.

11. Colaert, N., Degroeve, S., Helsens, K., & Martens, L. (2011). Analysis of the resolution limitations of peptide identification algorithms. Journal of Proteome Research, 10 (12), 5555–5561.
12. Coordinators, N. R. (2016). Database resources of the National Center for Biotechnology Information. *Nucleic Acids Research, 44* (D1), D7–D19.
13. Cox, J., & Mann, M. (2008). MaxQuant enables high peptide identification rates, individualized p.p.b.-range mass accuracies and proteome-wide protein quantification. *Nature Biotechnology, 26* (12), 1367–1372.
14. Cox, J., Neuhauser, N., Michalski, A., Scheltema, R. A., Olsen, J. V., & Mann, M. (2011). Andromeda: A peptide search engine integrated into the MaxQuant environment. *Journal of Proteome Research, 10* (4), 1794–1805.
15. Craig, R., & Beavis, R. C. (2004). TANDEM: Matching proteins with tandem mass spectra. *Bioinformatics, 20* (9), 1466–1467.
16. Craig, R., Cortens, J. P., & Beavis, R. C. (2004). Open source system for analyzing, validating, and storing protein identification data. *Journal of Proteome Research, 3* (6), 1234–1242.
17. Crappe, J., Ndah, E., Koch, A., Steyaert, S., Gawron, D., De Keulenaer, S., De Meester, E., De Meyer, T., Van Criekinge, W., Van Damme, P., & Menschaert, G. (2015). PROTEOFORMER: Deep proteome coverage through ribosome profiling and MS integration. *Nucleic Acids Research, 43* (5), e29.
18. Croft, D., O'Kelly, G., Wu, G., Haw, R., Gillespie, M., Matthews, L., Caudy, M., Garapati, P., Gopinath, G., Jassal, B., Jupe, S., Kalatskaya, I., Mahajan, S., May, B., Ndegwa, N., Schmidt, E., Shamovsky, V., Yung, C., Birney, E., Hermjakob, H., D'Eustachio, P., & Stein, L. (2011). Reactome: A database of reactions, pathways and biological processes. *Nucleic Acids Research, 39* (Database issue), D691–D697.
19. Desiere, F., Deutsch, E. W., King, N. L., Nesvizhskii, I., Mallick, P., Eng, J., Chen, S., Eddes, J., Loevenich, S. N., & Aebersold, R. (2006). The PeptideAtlas project. *Nucleic Acids Research, 34* (Database issue), D655–D658.
20. Deutsch, E. W., Mendoza, L., Shteynberg, D., Farrah, T., Lam, H.,

Tasman, N., Sun, Z., Nilsson, E., Pratt, B., Prazen, B., Eng, J. K., Martin, D. B., Nesvizhskii, A. I., & Aebersold, R. (2010). A guided tour of the trans-proteomic pipeline. *Proteomics, 10* (6), 1150– 1159.

21. Diament, B. J., & Noble, W. S. (2011). Faster SEQUEST searching for peptide identification from tandem mass spectra. *Journal of Proteome Research, 10* (9), 3871–3879.
22. Domon, B., & Aebersold, R. (2006). Mass spectrometry and protein analysis. *Science, 312* (5771), 212–217.
23. Dorfer, V., Pichler, P., Stranzl, T., Stadlmann, J., Taus, T., Winkler, S., & Mechtler, K. (2014). MS Amanda, a universal identification algorithm optimized for high accuracy tandem mass spectra. *Journal of Proteome Research, 13* (8), 3679–3684. doi:
24. Elias, J. E., & Gygi, S. P. (2010). Target-decoy search strategy for mass spectrometry-based proteomics. *Methods in Molecular Biology, 604*, 55–71.
25. Eng, J. K., Jahan, T. A., & Hoopmann, M. R. (2013). Comet: An open-source MS/MS sequence database search tool. *Proteomics, 13* (1), 22–24.
26. Everett, L. J., Bierl, C., & Master, S. R. (2010). Unbiased statistical analysis for multi-stage proteomic search strategies. *Journal of Proteome Research, 9* (2), 700–707.
27. Fan, J., Saha, S., Barker, G., Heesom, K. J., Ghali, F., Jones, A. R., Matthews, D. A., & Bessant, C. (2015). Galaxy integrated Omics: Web-based standards compliant workflows for proteomics informed by transcriptomics. *Molecular & Cellular Proteomics, 14* (11), 3087–3093.
28. French, W. R., Zimmerman, L. J., Schilling, B., Gibson, W., Miller, C. A., Townsend, R. R., Sherrod, S. D., Goodwin, C. R., McLean, J. A., & Tabb, D. L. (2015). Wavelet-based peak detection and a new charge inference procedure for MS/MS implemented in ProteoWizard's msConvert. *Journal of Proteome Research, 14* (2), 1299–1307.
29. Giardine, B., Riemer, C., Hardison, R. C., Burhans, R., Elnitski, L., Shah, P., Zhang, Y., Blankenberg, D., Albert, I., Taylor, J., Miller, W., Kent, W. J., & Nekrutenko, A. (2005). Galaxy: A platform for interactive large-scale genome analysis. Genome Research, 15 (10), 1451–1455.
30. Griss, J., Foster, J. M., Hermjakob, H., & Vizcaino, J. A. (2013). PRIDE

Cluster: Building a consensus of proteomics data. *Nature Methods, 10* (2), 95–96.

31. Helsens, K., Timmerman, E., Vandekerckhove, J., Gevaert, K., & Martens, L. (2008). Peptizer, a tool for assessing false positive peptide identifications and manually validating selected results. *Molecular & Cellular Proteomics, 7* (12), 2364–2372.
32. Henry, V. J., Bandrowski, A. E., Pepin, A. S., Gonzalez,J., & Desfeux, A. (2014). OMICtools: An informative directory for multi-omic data analysis. Database (Oxford). 2014 Jul 14
33. Hubbard, T., Barker, D., Birney, E., Cameron, G., Chen, Y., Clark, L., et al. (2002). The Ensemble genome database project. Nucleic *Acids Research, 30* (1), 38–41.
34. Hulstaert, N., Reisinger, F., Rameseder, J., Barsnes, H., Vizcaino, J. A., & Martens, L. (2013). Pride-asap: Automatic fragment ion annotation of identified PRIDE spectra. *Journal of Proteomics, 95,* 89–92.
35. Jaffe, J. D., Berg, H. C., & Church, G. M. (2004). Proteogenomic mapping as a complementary method to perform genome annotation. *Proteomics, 4* (1), 59–77.
36. Jagtap, P. D., Johnson, J. E., Onsongo, G., Sadler, F. W., Murray, K., Wang, Y., Shenykman, G. M., Bandhakavi, S., Smith, L. M., & Griffin, T. J. (2014). Flexible and accessible workfl ows for improved proteogenomic analysis using the Galaxy framework. *Journal of Proteome Research, 13* (12), 5898–5908.
37. Keller, A., Nesvizhskii, A. I., Kolker, E., & Aebersold, R. (2002). Empirical statistical model to estimate the accuracy of peptide identifications made by MS/MS and database search. *Analytical Chemistry, 74* (20), 5383–5392.
38. Kessner, D., Chambers, M., Burke, R., Agus, D., & Mallick, P. (2008). ProteoWizard: Open source software for rapid proteomics tools development. *Bioinformatics, 24* (21), 2534–2536.
39. Kim, S., & Pevzner, P. A. (2014). MS-GF+ makes progress towards a universal database search tool for proteomics. *Nature Communications, 5,* 5277.
40. Knudsen, G. M., & Chalkley, R. J. (2011). The effect of using an inappropriate protein database for proteomic data analysis. *PLoS One, 6* (6), e20873.
41. Kohlbacher, O., Reinert, K., Gropl, C., Lange, E., Pfeifer, N., Schulz-

Trieglaff, O., & Sturm, M. (2007). TOPP– the OpenMS proteomics pipeline. *Bioinformatics, 23* (2), e191–e197.

42. Lange, E., Gropl, C., Reinert, K., Kohlbacher, O., & Hildebrandt, A. (2006). High-accuracy peak picking of proteomics data using wavelet techniques. In *Pacific Symposium on Biocomputing Pacific Symposium on Biocomputing*, pp. 243–254.
43. Ma, K., Vitek, O., & Nesvizhskii, A. I. (2012). A statistical model-building perspective to identification of MS/ MS spectra with PeptideProphet. *BMC Bioinformatics, 13* (Suppl 16), S1.
44. Martens, L., Hermjakob, H., Jones, P., Adamski, M., Taylor, C., States, D., Gevaert, K., Vandekerckhove, J., & Apweiler, R. (2005a). PRIDE: The proteomics identifications database. *Proteomics, 5* (13), 3537–3545.
45. Martens, L., Nesvizhskii, A. I., Hermjakob, H., Adamski, M., Omenn, G. S., Vandekerckhove, J., & Gevaert, K. (2005b). Do we want our data raw? Including binary mass spectrometry data in public proteomics data repositories. *Proteomics, 5* (13), 3501–3505.
46. Menschaert, G., & Fenyo, D. (2015). Proteogenomics from a bioinformatics angle: A growing field. Mass Spectrom Rev. 2017 Sep;36(5):584-599.
47. Muth, T., Benndorf, D., Reichl, U., Rapp, E., & Martens, L. (2013a). Searching for a needle in a stack of needles: Challenges in metaproteomics data analysis. *Molecular BioSystems, 9* (4), 578–585.
48. Muth, T., Peters, J., Blackburn, J., Rapp, E., & Martens, L. (2013b). ProteoCloud: A full-featured open source proteomics cloud computing pipeline. *Journal of Proteomics, 88,* 104–108.
49. Nesvizhskii, A. I. (2010). A survey of computational methods and error rate estimation procedures for peptide and protein identification in shotgun proteomics. *Journal of Proteomics, 73* (11), 2092–2123.
50. Nesvizhskii, A. I. (2014). Proteogenomics: Concepts, applications and computational strategies. *Nature Methods, 11* (11), 1114–1125.
51. Nesvizhskii, A. I., & Aebersold, R. (2005). Interpretation of shotgun proteomic data: The protein inference problem. *Molecular & Cellular Proteomics, 4* (10), 1419–1440.
52. Noble, W. S. (2015). Mass spectrometrists should search only for peptides they care about. *Nature Methods, 12* (7), 605–608.
53. Olexiouk, V., Crappe, J., Verbruggen, S., Verhegen, K., Martens, L.,

& Menschaert, G. (2016). sORFs.org: A repository of small ORFs identified by ribosome profiling. *Nucleic Acids Research, 44* (D1), D324–D329.

54. Pemovska, T., Kontro, M., Yadav, B., Edgren, H., Eldfors, S., Szwajda, A., et al., (2013). Individualized systems medicine strategy to tailor treatments for patients with chemorefractory acute myeloid leukemia. *Cancer Discovery, 3* (12), 1416– 1429.
55. Pruitt, K. D., Tatusova, T., & Maglott, D. R. (2005). NCBI Reference Sequence (RefSeq): A curated nonredundant sequence database of genomes, transcripts and proteins. *Nucleic Acids Research, 33* (Database issue), D501–D504.
56. Risk, B. A., Spitzer, W. J., & Giddings, M. C. (2013). Peppy: Proteogenomic search software. *Journal of Proteome Research, 12* (6), 3019–3025.
57. Shanmugam, A. K., & Nesvizhskii, A. I. (2015). Effective leveraging of targeted search spaces for improving peptide identification in Tandem Mass Spectrometry based proteomics. *Journal of Proteome Research, 14* (12), 5169–5178.
58. Sheynkman, G. M., Johnson, J. E., Jagtap, P. D., Shortreed, M. R., Onsongo, G., Frey, B. L., Griffin, T. J., & Smith, L. M. (2014). Using Galaxy-P to leverage RNA-Seq for the discovery of novel protein variations. *BMC Genomics, 15*, 703.
59. Sussman, J. L., Lin, D., Jiang, J., Manning, N. O., Prilusky, J., Ritter, O., & Abola, E. E. (1998). Protein Data Bank (PDB): Database of three-dimensional structural information of biological macromolecules. Acta Crystallographica. Section D: Biological Crystallography, 54 (Pt 6 Pt 1), 1078–1084.
60. Tabb, D. L., Fernando, C. G., & Chambers, M. C. (2007). MyriMatch: Highly accurate tandem mass spectral peptide identification by multivariate hypergeometric analysis. *Journal of Proteome Research, 6* (2), 654–661.
61. Tateno, Y., Imanishi, T., Miyazaki, S., Fukami-Kobayashi, K., Saitou, N., Sugawara, H., & Gojobori, T. (2002). DNA Data Bank of Japan (DDBJ) for genome scale research in life science. *Nucleic Acids Research, 30* (1), 27–30.
62. Trudgian, D. C., & Mirzaei, H. (2012). Cloud CPFP: A shotgun proteomics data analysis pipeline using cloud and high performance

computing. Journal of Proteome Research, 11 (12), 6282–6290.

63. Vaudel, M., Barsnes, H., Berven, F. S., Sickmann, A., & Martens, L. (2011a). SearchGUI: An open-source graphical user interface for simultaneous OMSSA and X!Tandem searches. *Proteomics, 11* (5), 996–999.

64. Vaudel, M., Burkhart, J. M., Breiter, D., Zahedi, R. P., Sickmann, A., & Martens, L. (2012a). A complex standard for protein identification, designed by evolution. *Journal of Proteome Research, 11* (10), 5065–5071.

65. Vaudel, M., Burkhart, J. M., Sickmann, A., Martens, L., & Zahedi, R. P. (2011b). Peptide identification quality control. *Proteomics, 11* (10), 2105–2114.

66. Vaudel, M., Burkhart, J. M., Zahedi, R. P., Oveland, E., Berven, F. S., Sickmann, A., Martens, L., & Barsnes, H. (2015). PeptideShaker enables reanalysis of MS-derived proteomics data sets. *Nature Biotechnology, 33* (1), 22–24.

67. Vaudel, M., Sickmann, A., & Martens, L. (2010). Peptide and protein quantification: A map of the minefield. *Proteomics, 10* (4), 650–670.

68. Vaudel, M., Sickmann, A., & Martens, L. (2012b). Current methods for global proteome identification. *Expert Review of Proteomics, 9* (5), 519–532. Vaudel, M., Sickmann, A., & Martens, L. (2014a). Introduction to opportunities and pitfalls in functional mass spectrometry based proteomics. Biochimica et biophysica acta, 1844 (1 Pt A), 12–20.

69. Vaudel, M., Venne, A. S., Berven, F. S., Zahedi, R. P., Martens, L., & Barsnes, H. (2014b). Shedding light on black boxes in protein identification. *Proteomics, 14* (9), 1001–1005.

70. Vaudel, M., Verheggen, K., Csordas, A., Raeder, H., Berven, F. S., Martens, L., Vizcaino, J. A., & Barsnes, H. (2016). Exploring the potential of public proteomics data. *Proteomics, 16* (2), 214–225.

71. Verheggen, K., Barsnes, H., & Martens, L. (2014). Distributed computing and data storage in proteomics: Many hands make light work, and a stronger memory. *Proteomics, 14* (4–5), 367–377.

72. Verheggen, K., Maddelein, D., Hulstaert, N., Martens, L., Barsnes, H., & Vaudel, M. (2015). Pladipus enables universal distributed computing in proteomics bioinformatics. *Journal of Proteome Research.*

73. Vizcaino, J. A., Deutsch, E. W., Wang, R., Csordas, A., Reisinger, F., Rios, D., et al., (2014). ProteomeXchange provides globally

coordinated proteomics data submission and dissemination. *Nature Biotechnology, 32* (3), 223–226.

74. Vizcaino, J. A., Mueller, M., Hermjakob, H., & Martens, L. (2009). Charting online OMICS resources: A navigational chart for clinical researchers. *Proteomics Clinical Applications, 3* (1), 18–29.
75. Volders, P. J., Verheggen, K., Menschaert, G., Vandepoele, K., Martens, L., Vandesompele, J., & Mestdagh, P. (2015). An update on LNCipedia: A database for annotated human lncRNA sequences. *Nucleic Acids Research, 43* (Database issue), D174–D180.
76. Whiteaker, J. R., Halusa, G. N., Hoofnagle, A. N., Sharma, V., MacLean, B., Yan, P., Wrobel, J. A., Kennedy, J., Mani, D. R., Zimmerman, L. J., Meyer, M. R., Mesri, M., Rodriguez, H., Clinical Proteomic Tumor Analysis, C., & Paulovich, A. G. (2014). CPTAC assay portal: A repository of targeted proteomic assays. *Nature Methods, 11* (7), 703–704.
77. Yates, A., Akanni, W., Amode, M. R., Barrell, D., Billis, K., Carvalho-Silva, D., et al. (2016). Ensemble 2016. *Nucleic Acids Research, 44* (D1), D710–D716.
78. Zhang, B., Wang, J., Wang, X., Zhu, J., Liu, Q., Shi, Z., et al. (2014). Proteogenomic characterization of human colon and rectal cancer. *Nature, 513* (7518), 382–387.

CHAPTER 4

PROTEOGENOMICS SOFTWARES

CONTENTS

The newly introduced proteogenomics approach allows for an integrative incorporation of data extrapolated from the genomic and transcriptomics data and the possibility of wide-scale peptide detection (Armengaud, 2009). The recognized novel peptides through this proteogenomics approach enhance identification of novel genes, intronic and annotated untranslated regions and splice variants. In order to maximize the benefits from such promising approach, the creation of a precise searching database for a better representation of the new proteoforms becomes mandatory.

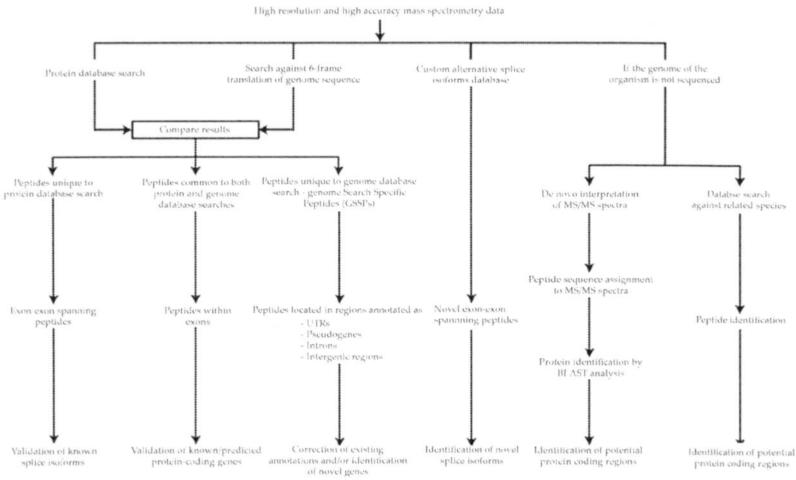

Figure 4.1. Highlights of proteogenomic approaches and associated potential discoveries.

Accordingly, a myriad of software pipelines and tools have been assembled. The basic units of proteogenomics analysis comprise the creation of a database, the identification of a peptide, mapping the genome for the detected peptides and giving inference on a corrected or a novel gene model. Accordingly, the current working tools either work as an end-to-end solution for -proteogeomic data analysis or work through the creation of the custom database. Such tools and pipelines allow for a reduction of potential errors and technical difficulties associated with the integration of this new approach. Meanwhile, the ease of use of these software programs allows for better expansion of the approach among molecular scientists.

4.1. PROTEOGENOMICS SOFTWARE

A number of the recently developed tools provide only a portion of the proteogenomic analysis, while few counts of pipelines deliver a complete proteogenomics workflow system. An illustration of these tools is provided in the Table 4.1.

Table 4.1. Tools and Pipelines of Proteogenomics

Tools and Pipelines	Description	Reference
CustomProDB	This is an R package that enhances the construction of custom proteogenomic databases by incorporating SNP information from RNA-Seq data or from a common variant call format (vcf) file.	Wang and Zhang, 2013
SpliceDB	Software that allows for the construction of the highly sensitive yet compact splice graph database. It is delivered in a FASTA format which can be searched by any tools of the peptide identification.	Burset et al., 2001
MSProGene	An application that allows for the construction of a sample specific search database from RNA-Seq data through the aid of network information of peptide sharing among each database entry	Zickmann and Renard, 2015
The Proteogenomic Mapping Tool	This tool allows for a better mapping of peptides back to the genome in an effective and quick way.	Sanders et al., 2011
SpliceVista	SpliceVista is a Python package that allows for mapping of the identified peptides on all of the known protein splice-variants. It also enhances the merged visualization of proteomics data with transcript information.	Zhu et al., 2014
dasHPPboard	This is a HUPO platform of endorsed data integration. It permits the analysis and visualization of various omics datasets comprising proteomics	Tabas-Madrid et al., 2015
VESPA	This is a JAVA based application that allows for the integrated visualization of proteomic and transcriptomic datasets in proteogenomic framework	Peterson et al., 2012
iPiG	This tool integrates the peptide identification into genome browser and therefore, allows for the concurrent analysis of multiple omics datasets	Kuhring and Renard, 2012
PGx	This is a modern tool that converts peptide identifications into browser extensible format (BED) containing genomic co-ordinates of sorts and can be visualized in genome browsers.	Askenazi et al., 2015
Genome Annotating Proteomic Pipeline (GAPP)	This pipeline was specifically designed to the human genome. This internet based application enhanced the annotation of numerous genes through the analysis of proteomic data that is publicly accessible. Unfortunately, this pipeline is no longer active for use	Shadforth et al., 2006

PepLine	It is unique software for genome annotation which is irrespective of database search method. Rather, it depends on a hybrid tag based search to characterize peptide tags and then maps and clusters these tags back to the genome to determine possible translated regions. It has only seen the limited application in proteogenomics research due to high-error rates of tag based peptide detection and genome mapping approach and the expected low sensitivity.	Ferro et al., 2008
Peppy	This is one of the initially established pipelines for proteogenomic analysis. It is an automated and fast framework for the quick search of MS data in front of the enormously large eukaryotic genome translated databases to discover new translated regions. The implementation of advanced computational methods in this tool makes proteogenomic searches usable on the simple desktop even for higher eukaryotic genomes which normally impose higher memory and highly computerized setup. In addition, it allows for a blind modification search to enhance detection of novel post-translation modifications which otherwise are very tough to identify by claasic proteomics searches. Despite these positive potentials, Peppy has limited eukaryotic analyses applicability as a large portion of new proteins in eukaryotes initiate from alternate splicing of transcripts which cannot be characterized in a genome translated search database	Risk et al., 2013
Enosi	This is a proteogenomics pipeline comprising two functional parts. First, SpliceDB tool (Burset et al., 2001) which is used to generate a comprehensive and compact database of splice junctions from RNA-Seq reads. Such fasta formatted splice graph database is then searched with MS data using MS-GF+ search engine (Kim and Pevzner, 2014) which is a sensitive tool to detect additional peptides. To appraise novel proteogenomics findings comprising splice junctions, Enosi uses a probability scoring which put into account the odds of spectra and peptides allocated to the locus, the quality of the assigned peptide spectral matches and the shared mapping of the peptide. The *eventProb* probability score allows Enosi to rank and filter the proteogenomic events depending on their confidence. Furthermore, the framework can employ *ab initio* gene predictions and RNASeq information to evaluate the boundaries of alternate gene models which accommodate the characterized novel peptides. In addition, Enosi pipeline is fully automated software and utilizes multi-threading to speed up the searches of the MS data.	Castellana et al., 2014

PGTools	is an end to end tool which assimilates multiple components of proteogenomic analysis. This is an open software suite which delivers fully automated searches along with the meta-analysis and visualization of new results. It permits the search against various custom databases, *e.g.*, databases comprising translated entries from transcripts, UTRs, non-coding genes, splice junctions, six frame translated genome and somatic variations. By allowing searches in front of cancer specific disparities from the COSMIC database and fusion proteins, PGTools also permits human cancer specific proteogenomic research work. Furthermore, its numerous search engine approach improves sensitivity by adding to the overall peptide detection process. Meanwhile, because of the variances in peptide detection confidence related to variable database sizes, result integration from these dissimilar databases provides novel challenges. In addition, the approach is short of the strength of individual or tissue specific proteogenomic searches as the comparable approaches of RNA-Seq data	Nagaraj et al., 2015
ProteoAnnotator	This is a modern, open source and highly powerful pipeline for proteogenomic mining from MS datasets. It presents one of the common problems of proteogenomic and proteomics research, which is the file format standard. The whole pipeline sustains and exports Proteomics Standards Initiative (PSI)- HUman Proteomics Organization (HUPO) –supported file formats. Proteoannotator also permits various database searches but basically depends on gene predictions. The Search of MS data in front of gene predictions is an exceptional approach for a novel sequenced genome basically because of the augmented sensitivity of peptide detection related to small search database in comparison to transcriptomics or genomic databases. In addition, the pipeline presents a "noncanonical gene model score" calculation which allows for the assignment of confidence values to new findings and therefore automated assessment of the quality of new discoveries. Additionally, it also introduces an automated framework which allows for the integration of multiple peptide search engines and comprehensive statistical algorithm. Despite it is highly effective for proteogenomically annotating new genomes, sample or individual based database searches are hard to use in this framework	Ghali et al., 2014

Integrated transcriptomic-proteomic pipeline (ITP)	This is a modern published pipeline and contains two analysis modules, each for proteomics and transcriptomics data. The transcriptomic analysis module employs Tuxedo suite of tools to assemble and align RNA-Seq results into transcripts by the utilization of the reference genome. The second module generates from the assembled transcripts a translated transcriptome database and then searches mass spectra in front of this database utilizing multiple search engines. Despite the pipeline is short of a whole automated structure for the public employment, the approach has numerous positive points. For example, the use of a reference genome guided transcriptome assembly yields an accurate transcript model for the discovered new peptides and therefore, proper reannotation of exon boundaries and coding splice variants are likely. Likewise, quantities of transcript isoforms may specify most possible protein coding isoform in spite of widespread peptide sharing among isoforms. It also permits the creation of individual or tissue specific search databases specifically suitable for the clinical research work. **EuGenoSui**te, enhance both the sensitivity and specificity of the detection of peptides. Recognized peptides are also transferred into gene transfer format (GTF) which can be simply integrated into a record of the genome browsers and therefore allowing simple visualization of new regions	Kumar et al., 2016b
PPLine	This is a Python grounded automated proteogenomics pipeline which allows for the integration of proteomics with transcriptome sequencing and exome sequencing technologies. Its chief intent is to explore variant novel peptides that result from single nucleotide polymorphism (SNP), insertions- deletions in the genomic DNA and that caused by alternative splicing. It also allows for the integration of numerous methods to accurately present SNPs from exome sequencing findings, aligns RNA-Seq results, assembles transcripts comprising splice junction isoforms from findings and then permits the search of proteomics data in front of variant peptide database. This complete software allows for tissue/sample specific database construction and therefore enhances the analysis of clinical proteogenomics.	Krasnov et al., 2015

GALAXY-P	This is one of the scarce web-grounded frameworks for proteogenomics. In spite of its web grounded utilization, it permits massive analysis of eukaryotic genomes with ease at all steps of analysis. It spreads the Galaxy bioinformatics framework for proteomics data analysis and lets the scientist to construct custom integrative analysis workflows. Basic workflows within Galaxy-P permit MS data format conversion, construction of proteogenomic databases from numerous web datasets, two step database search and statistical evaluation of recognized peptides, sequence correspondence searches of new discoveries, assessment of peptide-spectral matches by visualization and comprehensive genomic visualization of new peptides. Additionally, the Galaxy framework permits easy integration of numerous genomics and transcriptomics data analysis and with the Galaxy-P enhancement, integration of proteomics with other modern omics datasets becomes easy to use. For example, Sheynkman et al. (2014) constructed three analysis workflows which allow the search of proteomics data within Galaxy-P framework in front of splice variant database developed from RNA-Seq data and single amino acid polymorphism (SAP)	Jagtap et al., 2014
QUILTS	This is software to construct individual specific human proteogenomics search databases through the integration of splice variants, gene fusions, and SAP variations, to established protein sequences. Individual specific transcriptomics and genomic differences have been related to various diseases basically cancers and therefore, it should enhance clinical proteogenomic research work intended to discover disease specific variants. In spite of that, it is confined to human only and does not permit comparable analysis of other model organisms, used to investigate human disorders.	Zhang et al., 2009

With so many softwares and tools, one important question always arises: Which one is the best tool? Despite that there are no enough studies to compare the numerous pipelines intended for the study eukaryotic proteogenomics, one of the modern studies proposes that many of these tools and softwares are in fact complement the results of each other (Kumar et al., 2016b). This result could also be concluded because dissimilarities in their search database compositions bring complementary peptide detections. Despite the presence of various technical challenges to run integrated numerous proteogenomics pipelines on a wide scale proteomic study, this strategy would be highly beneficial and help to construct a comprehensive directory of new translation findings across the human and eukaryotic genome.

REFERENCES

1. Armengaud, J. (2009). A perfect genome annotation is within reach with the proteomics and genomics alliance. *Current Opinion in Microbiology, 12* (3), 292–300.
2. Askenazi, M., Ruggles, K. V., & Fenyo, D. (2015). PGx: Putting peptides to BED. *Journal of Proteome Research, 15* (3), 795–799.
3. Burset, M., Seledtsov, I. A., & Solovyev, V. V. (2001). SpliceDB: Database of canonical and non-canonical mammalian splice sites. *Nucleic Acids Research, 29* (1), 255–259.
4. CCastellana, N. E., Shen, Z., He, Y., Walley, J. W., Cassidy, C. J., Briggs, S. P., & Bafna, V. (2014). An automated proteogenomic method uses mass spectrometry to reveal novel genes in Zea mays. *Molecular & Cellular Proteomics, 13* (1), 157–167.
5. Ferro, M., Tardif, M., Reguer, E., Cahuzac, R., Bruley, C., Vermat, T., Nugues, E., Vigouroux, M., Vandenbrouck, Y., Garin, J., & Viari, A. (2008). PepLine: A software pipeline for high-throughput direct mapping of tandem mass spectrometry data on genomic sequences. *Journal of Proteome Research, 7* (5), 1873–1883.
6. Ghali, F., Krishna, R., Perkins, S., Collins, A., Xia, D., Wastling, J., & Jones, A. R. (2014). ProteoAnnotator – Open source proteogenomics annotation software supporting PSI standards. *Proteomics, 14* (23–24), 2731–2741.
7. Jagtap, P. D., Johnson, J. E., Onsongo, G., Sadler, F. W., Murray, K., Wang, Y., Shenykman, G. M., Bandhakavi, S., Smith, L. M., & Griffin, T. J. (2014). Flexible and accessible workfl ows for improved proteogenomic analysis using the galaxy framework. *Journal of Proteome Research, 13* (12), 5898–5908.
8. Kim, S., & Pevzner, P. A. (2014). MS-GF+ makes progress towards a universal database search tool for proteomics. *Nature Communications, 5,* 5277.
9. Krasnov, G. S., Dmitriev, A. A., Kudryavtseva, A. V., Shargunov, A. V., Karpov, D. S., Uroshlev, L. A., Melnikova, N. V., Blinov, V. M., Poverennaya, E. V., Archakov, A. I., Lisitsa, A. V., & Ponomarenko, E. A. (2015). PPLine: An automated pipeline for SNP, SAP, and splice variant detection in the context of proteogenomics. *Journal of Proteome Research, 14* (9), 3729–3737.

10. Kuhring, M., & Renard, B. Y. (2012). iPiG: Integrating peptide spectrum matches into genome browser visualizations. *PLoS One, 7* (12), e50246.
11. Kumar, D., Mondal, A. K., Kutum, R., & Dash, D. (2016a). Proteogenomics of rare taxonomic phyla: A prospective treasure trove of protein coding genes 2. *Proteomics, 16* (2), 226–240.
12. Kumar, D., Yadav, A. K., Jia, X., Mulvenna, J., & Dash, D. (2016b). Integrated transcriptomic-proteomic analysis using a proteogenomic workflow refines rat genome annotation. *Molecular Cellular Proteomics, 15* (1), 329–339.
13. Nagaraj, S. H., Waddell, N., Madugundu, A. K., Wood, S., Jones, A., Mandyam, R. A., Nones, K., Pearson, J. V., & Grimmond, S. M. (2015). PGTools: A software suite for proteogenomic data analysis and visualization. *Journal of Proteome Research, 14* (5), 2255– 2266.
14. Peterson, E. S., McCue, L. A., Schrimpe-Rutledge, A. C., Jensen, J. L., Walker, H., Kobold, M. A., Webb, S. R., Payne, S. H., Ansong, C., Adkins, J. N., Cannon, W. R., & Webb-Robertson, B. J. (2012). VESPA: Software to facilitate genomic annotation of prokaryotic organisms through integration of proteomic and transcriptomic data. *BMC Genomics, 13,* 131.
15. Risk, B. A., Spitzer, W. J., & Giddings, M. C. (2013). Peppy: Proteogenomic search software. *Journal of Proteome Research, 12* (6), 3019–3025.
16. Sanders, W. S., Wang, N., Bridges, S. M., Malone, B. M., Dandass, Y. S., McCarthy, F. M., Nanduri, B., Lawrence, M. L., & Burgess, S. C. (2011). The proteogenomic mapping tool. *BMC Bioinformatics, 12,* 115.
17. Shadforth, I., Xu, W., Crowther, D., & Bessant, C. (2006). GAPP: A fully automated software for the confident identification of human peptides from tandem mass spectra. *Journal of Proteome Research, 5* (10), 2849– 2852.
18. Sheynkman, G. M., Johnson, J. E., Jagtap, P. D., Shortreed, M. R., Onsongo, G., Frey, B. L., Griffin, T. J., & Smith, L. M. (2014). Using Galaxy-P to leverage RNA-Seq for the discovery of novel protein variations. *BMC Genomics, 15,* 703.
19. Tabas-Madrid, D., Alves-Cruzeiro, J., Segura, V., Guruceaga, E., Vialas, V., Prieto, G., Garcia, C., Corrales, F. J., Albar, J. P., & Pascual-

Montano, A. (2015). Proteogenomics dashboard for the human proteome project 1. *Journal of Proteome Research, 14* (9), 3738–3749.

20. Tanner, S., Shen, Z., Ng, J., Florea, L., Guigo, R., Briggs, S. P., & Bafna, V. (2007). Improving gene annotation using peptide mass spectrometry. *Genome Research, 17* (2), 231–239.
21. Wang, X., & Zhang, B. (2013). CustomProDB: An R package to generate customized protein databases from RNA-Seq data for proteomics search 1. *Bioinformatics, 29* (24), 3235–3237.
22. Zhang, G., Fenyo, D., & Neubert, T. A. (2009). Evaluation of the variation in sample preparation for comparative proteomics using stable isotope labeling by amino acids in cell culture. *Journal of Proteome Research, 8* (3), 1285–1292.
23. Zhu, Y., Hultin-Rosenberg, L., Forshed, J., Branca, R. M., Orre, L. M., & Lehtio, J. (2014). SpliceVista, a tool for splice variant identification and visualization in shotgun proteomics data. *Molecular Cell Proteomics, 13* (6), 1552–1562.
24. Zickmann, F., & Renard, B. Y. (2015). MSProGene: Integrative proteogenomics beyond six-frames and single nucleotide polymorphisms. *Bioinformatics, 31* (12), i106–i115.

CHAPTER 5

MUTANT PROTEOGENOMICS

CONTENTS

5.1. INTRODUCTION

The technical enhancements in the area of mass spectrometry (MS) based proteomics, particularly bottom-up proteomics, provides a solid system to test newly recognized open reading frames (ORFs) and genes (Pandey and Pevzner, 2014), which is in congruence with the Human Genome Organization (HUGO) project. HUPO, the Human Protein Organization, has analogously intended to achieve comparable objectives, comprising the accomplishment of the human protein catalogue (Omenn et al., 2015). To enhance the recognition of proteins, databases have been established and sustained, adding and enhancing protein sequences in a continuous way in these repositories.

The most commonly employed protein databases (e.g neXtProt and UniProt/SwissProt) were planned to comprise the most frequent forms of human proteins, therefore the description of canonical or consensus protein sequences were defined. As a reason of the communication between proteomics and genomic research, these protein databases were protracted with established splicing isoforms (alternative splicing variants or ASVs), augmenting the count of inputs by a factor of 2 to 3 (for example, the neXtProt has 20,066 consensus and 21,932 ASVs). In addition, a major consideration has been assumed to post-translational modifications (PTMs), such as ubiquitination, phosphorylation, glycosylation, etc. Data extrapolated from PTMs elaborates a large amount of new data about proteoforms with numerous functions that could be employable in explanation of the progression of disease (Nørregaard Jensen, 2004).

Remarkably, however, minor consideration has been revealed towards mutant proteins, despite they denote a level of variability of molecular forms between PTMs and ASVs. Particularly, single amino acid variants (SAAVs) are under consideration, This is due to 1. Genetic data/ information is frequently available 2. They may be the definitive markers as their exceptional sequence may change the function 3. Finally, they may complicate the quantification of given proteins.

The significance of discovery non-synonymous single nucleotide polymorphisms (nsSNPs) does not only lie in the finding of dissimilarities in the amino acid sequences that have functional consequences but also to give information with regard to the genetic, and probably phenotypic, the variance within the samples (Salisbury et al., 2003). Secondary confirmation by sequencing of corresponding genomic DNA has affirmed the existence of the predicted single nucleotide polymorphisms (SNPs) in 8 out of 10 SNP-peptides. In a research paper, Bunger et al. emphasized the practicality of inferring unassigned spectra as polymorphisms (Bunger et al., 2007).

Despite DNA genotyping scans have somehow the huge usefulness in identifying the haplotype structure on a genome-wide scale; proteins are a main functional constituent of many disease developments and progressions. Hence, information received from being capable to constantly observe SNP products in proteomic data permits an additional functional inference to be allocated to the expressed alleles. In relation, the usefulness of the detection of expressed SNPs in proteomics assays will help the integration of protein profiling with genome data. This analysis will disclose differential allelic expressions that can be interrelated to individual phenotypic variation. The modern concern in differential allelic expression has been motivated by the finding that about 50% of human heterozygous alleles are differentially expressed by a factor of two or more.

5.2. THEORETICAL CONSIDERATIONS

Gene mutation, by definition, is a permanent change of the DNA region defined as a gene. A gene mutation can comprise a number of alterations that can be categorized as a single nucleotide variation, changes in a longer sequence, all the way up to a large change in a segment of a chromosome that comprises various genes. In contrast, from a biological base, gene mutations can be distributed into two sorts: acquired (somatic) and hereditary (germline), subject to whether germ or somatic cells are the frames of the mutated DNA. Other than the most frequent form of genes (wild type), almost all genes have characteristic variations due to the great frequency of mutations. If a genetic change takes place in more than 1% of the whole population, such a variation is named a polymorphism. SNPs are common in the human genome: an account of 3.1 million SNPs have been established by the International HapMap Consortium (Frazer et al., 2007).

The effects of a SNP in a gene regarding the production of the protein can be identical, nonsense, and non-synonymous, resulting in unchanged (i.e., fully functional) or truncated and mutant protein sequences, respectively (Figure 5.1). Unchanging or the non-synonymous SNPs definitely add to the complexity of the proteome but can also give important insight into genetic changeability when matching individuals in a population.

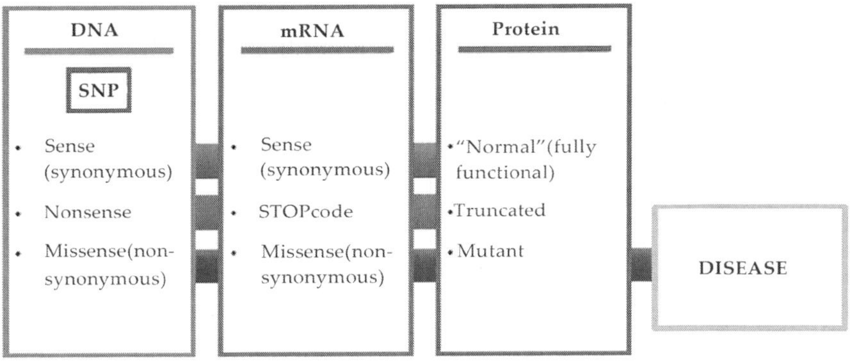

Figure 5.1. Schematic illustration of single nucleotide polymorphisms in the coding regions of DNA.

Whereas several polymorphisms are harmless and accountable for frequent variations in humans, some of them can add to the risk of the progression of the disease. It is acknowledged that somatic mutations can add to cancer development and their accumulation in the mitochondrial DNA is lined to an augmented risk of some age-related disorders, such as neurodegenerative and cardiovascular diseases (Taylor and Turnbull, 2005).

Variations can add to an augmented possibility to develop some diseases on the basis of the genetic map of an individual, which is commonly observed as genetic susceptibility or predisposition. For example, we can refer single amino acid variants of the BRCA1 and BRCA2 genes that can specify a significantly augmented risk of developing ovarian and breast cancer. BRCA1 and BRCA2 are human genes that yield tumor suppressor proteins, which aid to repair damaged DNA. Upon their mutation, their products are not made or do not function in a correct way, which leads to impaired DNA repair, which can subsequently result in an augmented possibility to develop extra genetic changes in the cells and ultimately lead to cancer. The comprehension of the underlying biological mechanisms at the molecular stage has leaded to an enhanced clinical diagnosis, observing these mutations in risk groups of patients. The possibility of elevated risk of breast cancer development may fluctuate as mutations in other genes, like BARD1 and BRIP1, are classically also linked to the disease.

In addition, protein aggregation has been identified in the development of neurological diseases, such as Alzheimer's disease, Jacob-Krautzfeld disease (associated with prions) and other amyloid diseases (e.g., familial amyloidotic polyneuropathy). A single amino acid alteration may importantly affect protein association and dissociation. For example, in the case of the familial amyloidotic polyneuropathy (FAP), transthyrein monomer units can combine into fibrils that can result in death upon deposition in the lungs and heart. The SAAVs of transthyrein, p.V30M and p.L55P, ease the dissociation of aggregates, whereas the p.T119M proteoform impedes tetramer dissociation (Hammarström et al., 2003). The chief cause for this inhibition is that position 119 is positioned at the dimer interface. Nevertheless, low-molecular-weight compounds could be developed to competently constrain transthyrein aggregation through the binding to the tetrameric form. Generally, the precise structure of mutant proteoforms helps to the development of efficient drugs.

The Pharmaceuticals and Medical Devices Agency in Japan was the leading agency in the world to approve personalized therapy in the event of non-small cell lung cancer, endorsing the application of tyrosine kinase inhibitors such as erlotinib and gefitinib. The epidermal growth factor receptor (EGFR) is overexpressed in the cells of definite types of human carcinomas (e.g., in breast and lung cancers), which results in an unsuitable activation of the anti-apoptotic Ras signaling cascade, ultimately leading to uncontrolled cell proliferation. It was revealed that a mutation in the EGFR tyrosine kinase domain is accountable for the activation of the anti-apoptotic pathways in non-small cell lung cancers (Sordella et al., 2004). These somatic mutations occur more frequently in lung adenocarcinomas of individuals of Asian descent, non-smokers, and women, making gefitinib/erlotinib treatments remarkably efficient.

It must be regarded that the recognition of a specific gene as a disease marker is tough and it is even less possible that this one gene is accountable for disease development alone, inferring that a whole change in the genetic profile is estimated to be more indicative. Typically, variations in individual genes may somewhat augment the risk of disease development but the combination of mutations on multiple genes can lead to an important level of risk (Genetic risk assessment and BRCA mutation testing for breast and ovarian cancer susceptibility: recommendation statement, 2005). Currently, it is commonly agreed that the variations in several genes, with their minute individual effects, may cause the liability to several frequent diseases when combined, this comprises diseases like diabetes, cancer, obesity and cardiovascular disease.

In addition to the understanding of the multifaceted nature of several diseases, and to the enormously successful genetic surveys on a large count of samples in populations, it is significant to highlight the fact that genes do not possess biological function, only their products as expressed proteins. Hence, it is mandatory to study not only DNA variations but also their related mRNA and their coded proteins, which together can refer to expression activities. Additionally, the interaction between these principle "omics" areas appears to be unavoidable to fully disclose the biology of diseases, unluckily, the recent status of these fields does not always ease their integration.

5.3. METHODOLOGIES FOR DETECTION OF MUTANT PROTEINS

5.3.1. Mass Spectrometry Based Proteomics of Mutant Proteoforms

Mass spectrometry based proteomic platforms possess the ability to recognize a large number of proteins concurrently in a single analysis. Nevertheless, protein recognition by the most frequently applied bottom-up proteomics method largely depends on protein sequence databases. Because of the fact that such databases are containing consensus sequences, i.e., the most commonly detected proteoforms, mass spectra of SAAVs and several other modifications of the analyzed peptides will be unassigned. Principally, known ASVs can be recognized searching databases, like UniProt and neXtProt, if isoform specific peptides could be produced prior to MS analysis. Substitute digestion strategies, using a gathering of proteases for the production of specific peptide sequences may enhance ASV recognitions. Nevertheless, the recognitions of SAAVs in high quality MS/MS datasets can be simply searchable databases that comprise altered amino acid sequences (Nesvizhskii et al., 2006). Genomic data of nsSNPs can be translated and incorporated into protein sequence databases that can endorse genomic based data in present tandem mass spectra, and ultimately perceive novel proteoforms in biological samples.

5.3.1.1 Databases for Identification of SAAVs

The leading reported database planned for mutant protein recognitions was based on the International Protein Sequence (IPI) database that was largely employed, possessing about 70,000 human protein entries (Schandorff et al., 2007). The inclusion of SAAVs and single amino acid conflicts revealed in the SwissProt databases posed the problem of augmented sequence redundancy that could ultimately compromise the reliability of recognitions.

Hence, sequential alterations of proteins were attached to the consensus protein sequences with an addition of letter "J" (as a "spacer" to identify the extra information in entries) between each peptide with the mutation site flanked at both ends with a peptide. The resultant MSIPI database was accomplished with a decoy database and published together with each novel IPI release by EBI until its last version (v3.67) that possessed 87,062 human protein entries.

As a substitute, another protein database (K-SNPdb) was released with 125,622 tryptic peptides, which comprised sequences of the changed amino acids (Bunger et al., 2007). The creation was grounded on filtering the NCBI dbSNP for nsSNPs (Sherry et al., 2001) that surpassed ten million SNPs all through the entire human genome. The count of nsSNPs out of all coding region SNPs was about 65,000. Along with the filtering, manual allocation with the protein accession numbers, the position of nsSNPs and the amino acid alterations were derived from the NCBI protein database, constructing a fasta file with paired reference and alternate alleles. In order to enhance the false discovery rate, a decoy database (FalseSNPdb) with the same count of peptides and matching masses to the peptides of K-SNPdb was composed of IPI entries. This approach granted the recognition of 629 SAAVs, of which 36 were not existent in the protein databases of NCBI and IPI.

In a trial to gather comprehensive sequential data about human mutant proteins, particularly about those engaged in cancer, oncogenesis and tumor progression, a novel database (CanProVar) was constructed (Li et al., 2010). Data on protein variations from public resources, comprising the Human Proteome Initiative (HPI) (O'Donovan et al., 2001), the Directory of Somatic Mutations in Cancer (COSMIC) (Bamford et al., 2004), the Cancer Genome Atlas (TCGA) (Comprehensive genomic characterization defines human glioblastoma genes and core pathways, 2008), the Online Mendelian Inheritance in Man (OMIM) (Hamosh et al., 2005) and two large-scale cancer genome sequencing studies (Greenman et al., 2007; Sjoblom et al., 2006), were integrated with a special identification of cancer related disparities (crVAR). The ultimate version of CanProVar holds 41,541 non-cancer definite and 11,445 cancer related disparities (http:// bioinfo.vanderbuilt.edu/ canprovar/). Most significantly, this gathering of human mutant proteins is searchable on-line, providing enormously useful data related to cancer samples, supplementary data sources, publications accompanied by functional information on gene ontology annotations and interaction partners (Figure 5.2).

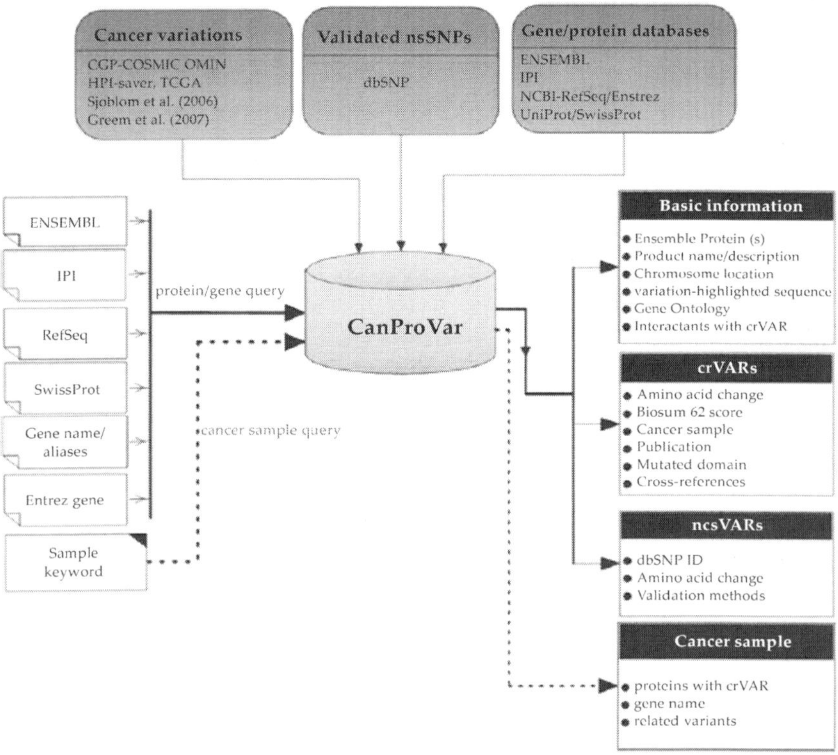

Figure 5.2. The system skeleton of CanProVar protein database that gives a large count of SAAV sequences (Li et al., 2010).

As a conclusion of the CanProVar project, an effective recognition workflow with numerous search engine options and a tool arranged for the correction of the false discovery rate (FDR), was presented and illustrated using CRC cell line data (Li et al., 2011). Remarkably, this downloadable MS-CanProVar database was finished with the Ensemble protein database (v53), 148 contaminant sequences and their reversed sequences as decoys (total searchable entry number is 290,440). Non-sense variations, missense variations, insertions and single amino acid deletions were all included in the database. To present the augmenting redundancy posed by the addition of mutant sequences, they were shortened to a tryptic peptide with the mutation site flanked by two peptides.

As a consequence, the inclusion of these novel peptides augmented the databased size by only 3.4%. The MS-CanProVar database represented the basis of a recent approach that combined it with the definite variant sequences in UniProt human polymorphisms (release 2011-12-14), leading to a total of 87,745 SAAVs (Song et al., 2014). The Swiss-CanSAAVs database comprises a total count of 161,7 downloadable entries (http://bioanalysis.icp.ac.cn/proteomics/Publications/SSAV/ SAAV-Database.htm) with diminished redundancy utilizing only tryptic mutant peptide sequences flanked by two additional peptides.

The Human Protein Mutant Database (HPMD), which is a customized database, was also produced through the extraction and combination of sequential information of known disease mutations from the Systematic Platform for Identifying Mutated Proteins (SysPIMP) (Xi et al., 2009), OMIM; the Protein Mutant Database (PMD) (Kawabata et al., 1999) and UniProt (Magrane and Consortium, 2011) (Figure 5.3) (Mathivanan et al., 2012). To enhance the FDR, sequence redundancy was diminished by limiting the peptide sequences to 101 amino acids with the mutation site in the center position (51st residue). This approach has a disadvantage as it runs an exclusion of the probable recognition of mutant proteins by a miss-cleaved tryptic peptide.

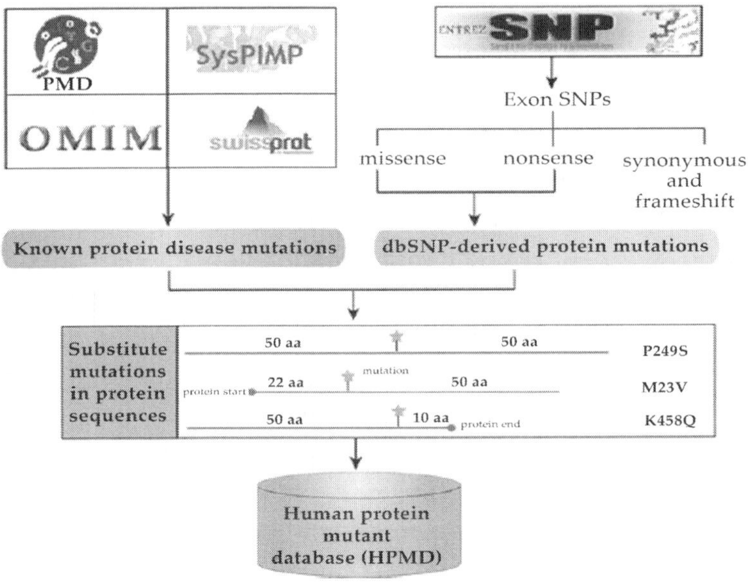

Figure 5.3. Human Protein Mutant Database (HPMD) construction for MS based protein mutation search (Mathivanan et al., 2012).

Despite the augmenting access to RNA-Seq data, the additional path has unlocked to create extended protein sequence databases, which is the translation of transcriptomic information to amino acids. The combination of next generation sequencing (NGS) and shotgun proteomics technologies has revealed to be an operative approach to reveal multilevel data and knowledge about cellular systems (Chen et al., 2012). To enhance translations of RNA-Seq sequences to protein levels, a smart bioinformatics tool was presented currently to allow for the creation of customized protein databases (Wang and Zhang, 2013). The R package of customProDB can simply create enhanced protein databases from RNA-Seq data with recognized single nucleotide variations, deletions, and short insertions as well as novel junctions between exons. The customProDB was an important and integrated component of the novel developed proteogenomics dashboard (dasHPPboard) envisioned to improve the protein mapping findings of HPP (Tabas-Madrid et al., 2015).

5.3.2. Approach for Identification of Mutant Proteins

The approach to recognizing new mutant proteoforms in biological samples was planned to utilize high quality shotgun proteomic tandem mass spectra for database search by current search engine algorithms (Lichti et al., 2015). The chief component of the strategy was the distinctive group of database entries that illustrates all SAAV sequences, translated from known genomic work (Ensembl). Utilizing a special custom made software tool, novel protein sequences were created to comprise a point mutation in each novel entry that therefore differed in a single amino acid from the consensus protein.

The mutant protein database (MuPdb) comprised 2.3 million SAAVs, with the exclusion of titin (Q8WZ42). The sequence redundancy was enormously decreased keeping the tryptic peptides with the mutation site flanked by two missed cleavages at both termini. The resultant in silico derived proteoforms were designated following the neXtProt nomenclature, comprising the access codes but also giving more information about the position and nature of the mutation (for example NX_P07288-SNP-L-132-I).

The MuPdb was put as a gathering of consensus (40,548 entries of UniProtKB) and mutant proteoform sequences of chromosome 19 (132,264 entries), together with 115 common contaminant sequences (cRAP) in standardized fasta format, so as to be utilized with nuermous search engines, comprising Proteome Discoverer, Mascot and PEAKS. To present the challenge that the large search space signifies, a custom decoy database was generated using PEAKS. In addition, manual validation of tandem mass spectra was excuted following blast searches for exceptionality. A primary recognition and validation of glioblastoma stem cells (GSC) exposed many SAAVs. Remarkably, a thoroughly checked mutation (p.T186R) of branched-chain aminotransferase 2 (BCAT2) was recognized (Lichti et al., 2015). This and other newly observed SAAVs in GSC samples were more checked at the transcript level and by SRM-assays planned for proper SAAV peptides.

This approach was widespread for the recognition of SAAVs in any biological sample as presented in Figure 5.4. After searches of high quality MS/MS data in the custom made a mutant database, the primary discoveries need verification. Presently, this step comprised both targeted proteomics and transcriptomic methods, a grouping of which is adequately powerful to deliver new biomarkers and drug targets in future applications.

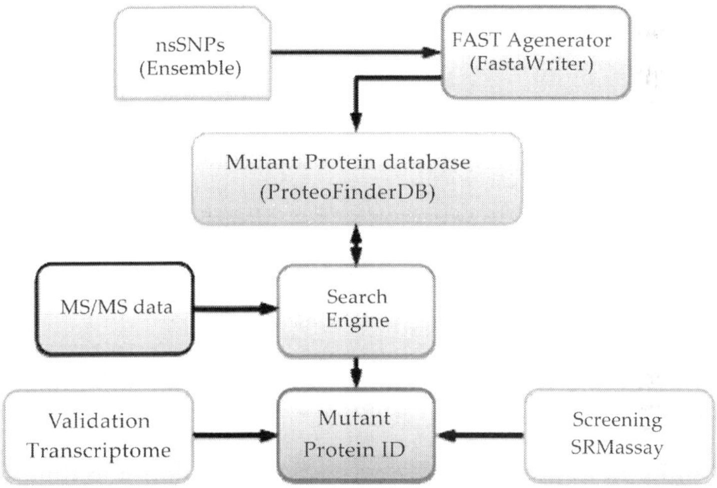

Figure 5.4. The general Scheme of recognition of SAAVs by tandem mass spectra searching a specialized protein database

The SRM-MS analysis for authentication of mutant proteoforms, directing toward the most probable tryptic peptides specific to mutation sites and their related wild type sequences, can also be excuted. Synthetic heavy isotope labeled peptides with relating sequences can be spiked into the clinical samples for unambiguous recognition of mutant proteoforms. Additionally, to give qualitative verification and quantification, the ratios between wild type and mutant forms can be assessed in heterozygous expression. In addition, RNA sequence analysis can be excuted with biological samples and the transcript data can be used to authenticate newly recognized mutant proteins. This method can yield considerable reference information and confirm SAAVs. Quantitative readout of RNA-Seq results can also be linked to known levels of expressed mutant proteins for authentication.

5.3.3. Targeted Proteomics of SAAVs

One of the most common mass spectrometry technologies, selected reaction monitoring (SRM), can be effectively applied to recognize and quantify definite peptides among the digested samples of complex mixtures (Feng and Picotti, 2016). Additionally, the SRM methodology is innately easy to multiplex, which allow for the initiation of numerous protein assays that provide high throughput and sensitivity. When practicing the stable isotope technology, uniformly 13 C- 15 N-labeled proteins can be assayed in blood plasma at levels of 100 ng/mL. Nevertheless, in some cases, extra enrichment steps are needed for the recognition of proteins existing at lower concentrations in human samples like plasma or serum.

The targeted enrichment, with or without antibodies, has been presented to enhance the detection sensitivity. Currently, numerous methods gathering immunoaffinity with SRM utilizing stable isotope peptides, such as immuno-SRM (Whiteaker et al., 2007; Whiteaker et al., 2011), SISCAPA-SRM (Anderson et al., 2004) and mass spectroÁ metric immunoassay (MSIA) (Lopez et al., 2010), have remarkably enhanced the limit of detection of low abundant protein biomarkers exising in plasma. Utilizing the MSIA approach, the lowermost detection level (LOQ) of plasma proteins is 16–31 pg/ml (Lopez et al., 2010). Immunoprecipitation (IP) is a logical approach to enrich mutant proteoforms in gathering with targeted proteomics techniques, such as SRM since antibodies can be produced against most proteins of interest and SRM-MS does not need absolute specificity for the antigens or for the relevant mutations.

In addition, IP can remove the most plentiful proteins, comprising serum albumin, cytoskeletal proteins, and immunoglobulins from biological samples (Anderson et al., 2004). As antibodies are not always accessible and can be expensive to advance, antibody-free enrichment of target proteins was currently illustrated for the quantitation of low plentiful plasma proteins at concentrations in the 50–100 pg/mL range (Shi et al., 2012).

5.3.4. Quantifications of SAAVs

The SRM technology provides definite and powerful quantifications of recognized proteins, targeting their described proteotypic peptides in biological samples. Using the high specificity of SRM-MS, multiplexing can be simply realized, providing a relatively high throughput approach. Nevertheless, the need for isotope labeled peptide standards for absolute or relative quantification may hinder this method, putting into consideration the difficulties to synthesize certain peptide sequences, the related costs may also be a restraint.

However, quantitative analysis of mutant proteoforms in disease progression studies can definitely give significant insights into the ratio of allele-specific gene expressions, which has been shown to be closely related to dissimilarities in individuals (Yan et al., 2002; Montgomery et al., 2010). It has been also illustrated that a SAAV of a single allele could actually be expressed in both of the homozygous and heterozygous manners (Végvári et al., 2013). The frequency of the mutant prostate specific antigen (PSA_ p.L132I) is in congruence with with population based genomic data, despite more large-scale and systematic studies are needed to comprehend how general this finding is (Figure 5.5).

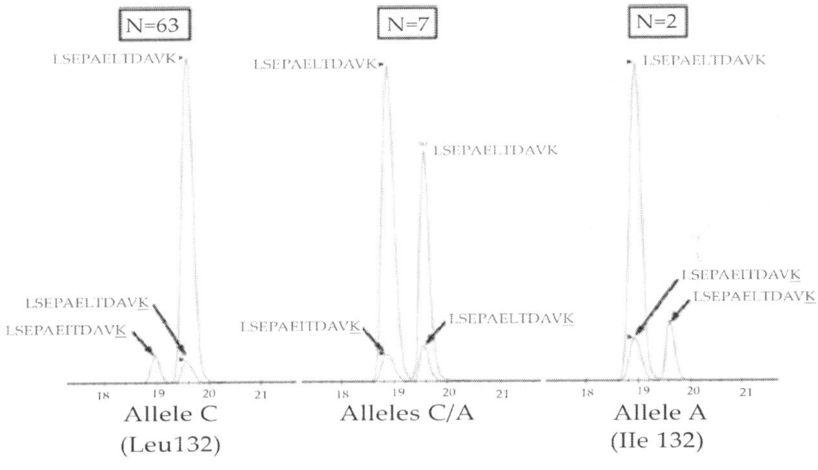

Figure 5.5. Identification of three possible combinations of allele expressions in examples of SRM-MS analyses (Végvári et al., 2013)

There are clues that disease progression may be observed by the level of mutant proteoforms in heterozygous expressions, which can enhance our appreciation of, for example, cancer biology. A boosting investigation was planned to use SRM based targeted proteomics to detect and quantify selected SAAVs in 290 clinical plasma samples taken from from Asian patients with both obesity and diabetes (Su et al., 2011). After the primary recognition of SAAVs in patient groups, comprising healthy controls, key proteotypic tryptic peptides were observed and their corresponding complement component proteins (C7, factor H, and C5) were checked by absolute quantification. The findings revealed that the homozygous expressions of factor H_ p.V62 and wild type C7_ p. T587 and were over represented in the selected Asian patient groups, while the C7_ p.T587P was remarkably higher in control samples.

In addition, no homozygous expression was revealed in any individuals, which is congruent well with past studies (Gimelbrant et al., 2007). Likewise, heterozygous expression profiles of these SAAVs were checked with adequate accuracy. This study has established the new concept of SRM-based quantitative analysis, which showed that the levels of heterozygous and homozygous SAAVs in patient populations have substantial linkage to certain disease traits.

An effective quantification method was established for the large-scale study of SAAVs in clinical samples (Song et al., 2014). The stable isotope dimethyl-labeling approach (Kovanich et al., 2012) could be utilized to quantify a total of 282 unique SAAV peptides in combined CID and HCD tandem mass spectra (Figure 5.6). As leucine (Leu) and isoleucine (Ile) are isobaric, SAAVs with changed Leu or Ile were omitted in the final results. The primary recognition of SAAVs was excuted using searches of mass spectra in front of a custom made protein database, comprising 87,745 amino acid variant sequences and 73,910 UniProt canonical protein entries (Swiss-CanSAAVs). Remarkably, the mutant sequences were shortened to a tryptic peptide with two missed cleavage at both flanks so as to decrease sequential redundancy and therefore enhance the false discovery rate (FDR). In addtion, the Swiss- CanSAAVs database was concatenated with the reversed sequences permitting the FDR analysis.

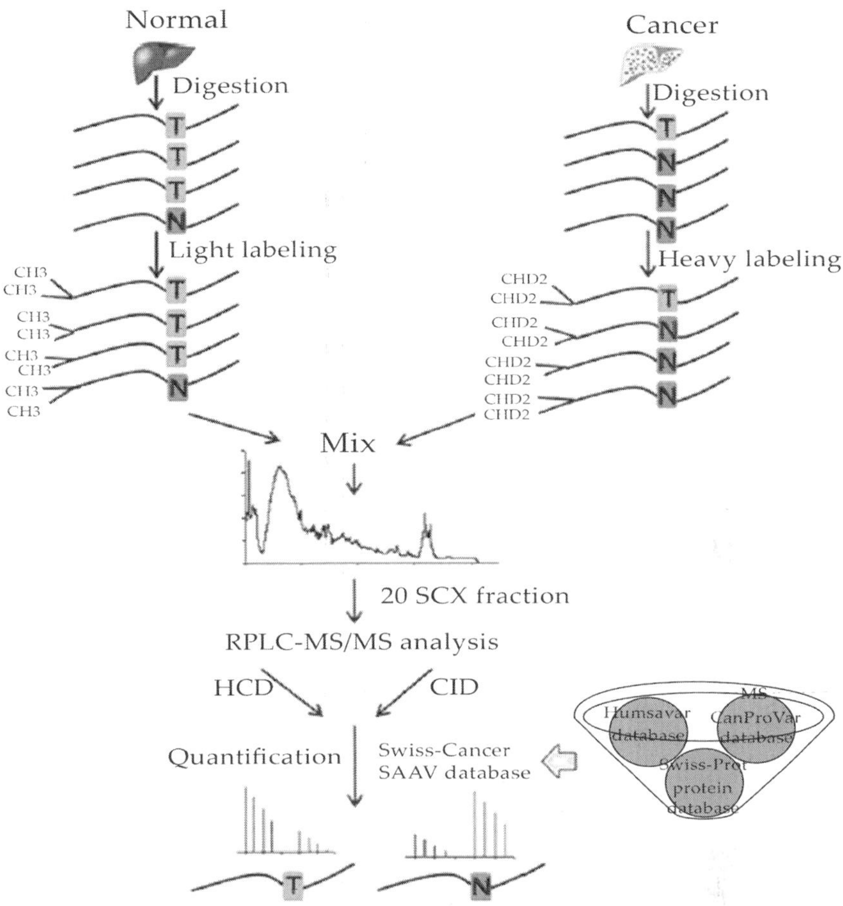

Figure 5.6. Analysis workflow for the wide-scale quantitative analysis of SAAVs between hepatocellular carcinoma and normal human liver tissues (Song et al., 2014).

5.4. BIOLOGICAL FINDINGS

Till now, little consideration has been given to the functional correlation between diseases and mutant proteins (Wang et al., 2011). Cancer research has currently revealed that solid tumors classically yield 20–100 mutant genes with non-synonymous variations, as identified through DNA-based sequencing studies (Wood et al., 2007). Current studies on some cancer types have revealed that mutant proteins can be correlated to disease and even be the initiator of disease (Bozic et al., 2010; Haber and Settleman,

2007). Depending on the occurrence and biological function of these mutant proteins, two groups were proposed: 1. "Drivers" that can drive and are accountable for tumor genesis 2. "Passengers" are not directly correlated with malignant differentiation (Haber and Settleman, 2007; Bozic et al., 2010).

Notably, the changed genetic codes of nsSNPs were revealed to be firmly correlated to the physiological and pathological traits of individuals (Sun et al., 2008). In addition, the ratio of allele-specific gene expressions in the heterozygous state is also correlated to numerous traits of individuals (Montgomery et al., 2010) and the quantitative association of the wild type/mutant proteins can also designate disease traits (Yan et al., 2002; Montgomery et al., 2010). As a consequence, both qualitative and quantitative data regarding the structures and the functional proteins of individuals with SAAVs are needed for a wide-ranging analysis.

Recognition and quantification of mutant proteoforms were excuted searching tandem mass spectra in front of a custom database that comprisedf 87,745 amino acid variant sequences and the whole entries of the canonical protein of UniProtKB (Song et al., 2014). This method was applied on profiling mutant proteomes in hepatocellular carcinoma (HCC) and healthy human tissue samples and leads to the identification of 282 unique SAAV sites. Notably, a remarkable increase of HIV-1 TATinteractive protein 2 (HTATIP2) p.S197R and carbamoyl phosphate synthase (CPS1) p.T1406N mutations were quantified in samples of HCC patients, which could be correlated to cancer progression (Song et al., 2014). Likewise, remarkable alteration of mutant proteomes was revealed in serum samples from pancreatic cancer patients and quantified using an isobaric labeling method (Nie et al., 2014). Consequently, a new biomarker panel was proposed, comprising thrombospondin-1 (THBS1), α-1-antichymotrypsin (AACT) and a mutant form of serotransferrin (TF_ p.V448I), that could distinguish pancreatic cancer patients from chronic pancreatitis and healthy control patients.

REFERENCES

1. Anderson, N. L., Anderson, N. G., Haines, L. R., Hardie, D. B., Olafson, R. W., & Pearson, T. W. (2004). Mass spectrometric quantitation of peptides and proteins using stable isotope standards and capture by anti- peptide antibodies (SISCAPA). *Journal of Proteome Research, 3* (2), 235–244.
2. Bamford, S., Dawson, E., Forbes, S., Clements, J., Pettett, R., Dogan, A., Flanagan, A., Teague, J., Futreal, P. A., Stratton, M. R., & Wooster, R. (2004). The COSMIC (Catalogue of Somatic Mutations in Cancer) database and website. *British Journal of Cancer, 91* (2), 355–358.
3. Bozic, I., Antal, T., Ohtsuki, H., Carter, H., Kim, D., Chen, S., Karchin, R., Kinzler, K. W., Vogelstein, B., & Nowak, M. A. (2010). Accumulation of driver and passenger mutations during tumor progression. *Proceedings of the National Academy of Sciences, 107* (43), 18545–18550.
4. Bunger, M. K., Cargile, B. J., Sevinsky, J. R., Deyanova, E., Na, Y., Hendrickson, R. C., & Stephenson, J. L. (2007). Detection and validation of non-synonymous coding SNPs from orthogonal analysis of shotgun proteomics data. *Journal of Proteome Research, 6*, 2331–2340.
5. Chen, R., Mias, G. I., Li-Pook-Than, J., Jiang, L., Lam, H. Y. K., Chen, R., et al. (2012). Personal omics profiling reveals dynamic molecular and medical phenotypes. *Cell, 148* (6), 1293–1307.
6. Comprehensive genomic characterization defines human glioblastoma genes and core pathways (2008). *Nature, 455* (7216), 1061–1068.
7. Feng, Y., & Picotti, P. (2016). Selected reaction monitoring to measure proteins of interest in complex samples: A practical guide. *Methods in Molecular Biology (Clifton, NJ), 139,* 43–56.
8. Frazer, K. A., Ballinger, D. G., Cox, D. R., Hinds, D. A., Stuve, L. L., Gibbs, R. A., et al. (2007) A second generation human haplotype map of over 3.1 million SNPs. *Nature 449* (7164), 851–861.
9. Genetic risk assessment and BRCA mutation testing for breast and ovarian cancer susceptibility: Recommendation statement. (2005). *Annals of Internal Medicine 143* (5),355–361.
10. Gimelbrant, A., Hutchinson, J. N., Thompson, B. R., & Chess, A. (2007). Widespread monoallelic expression on human autosomes. *Science, 318* (5853), 1136–1140.

11. Greenman, C., Stephens, P., Smith, R., Dalgliesh, G. L., Hunter, C., Bignell, G., et al. (2007). Patterns of somatic mutation in human cancer genomes. *Nature, 446* (7132), 153–158.
12. Haber, D. A., & Settleman, J. (2007). Cancer: Drivers and passengers. *Nature, 446* (7132), 145–146.
13. Hammarström, P., Wiseman, R. L., Powers, E. T., & Kelly, J. W. (2003). Prevention of transthyretin amyloid disease by changing protein misfolding energetics. *Science (New York, NY), 299* (5607), 713–716.
14. Hamosh, A., Scott, A. F., Amberger, J. S., Bocchini, C. A., & McKusick, V. A. (2005). Online Mendelian Inheritance in Man (OMIM), a knowledgebase of human genes and genetic disorders. *Nucleic Acids Research, 33* (Database issue), D514–517.
15. Kawabata, T., Ota, M., & Nishikawa, K. (1999). The protein mutant database. *Nucleic Acids Research, 27* (1), 355–357.
16. Kovanich, D., Cappadona, S., Raijmakers, R., Mohammed, S., Scholten, A., & Heck, A. J. (2012). Applications of stable isotope dimethyl labeling in quantitative proteomics. *Analytical and Bioanalytical Chemistry, 404* (4), 991–1009.
17. Li, J., Duncan, D. T., & Zhang, B. (2010). CanProVar: A human cancer proteome variation database. *Human Mutation, 31* (3), 219–228.
18. Li, J., Su, Z., Ma, Z. Q., Slebos, R. J. C., Halvey, P., Tabb, D. L., Liebler, D. C., Pao, W., & Zhang, B. (2011). A bioinformatics workfl ow for variant peptide detection in shotgun proteomics. *Molecular & Cellular Proteomics, 10* (5), M110.006536–M006110.006536.
19. Lichti, C. F., Mostovenko, E., Wadsworth, P. A., Lynch, G. C., Pettitt, B. M., Sulman, E. P., Wang, Q., Lang, F. F., Rezeli, M., Marko-Varga, G., Végvári, Á., & Nilsson, C. L. (2015). Systematic identification of single amino acid variants in Glioma stem-cell-derived chromosome 19 proteins. *Journal of Proteome Research, 14* (2), 778–786.
20. Lopez, M. F., Rezai, T., Sarracino, D. A., Prakash, A., Krastins, B., Athanas, M., Singh, R. J., Barnidge, D. R., Oran, P., Borges, C., & Nelson, R. W. (2010). Selected reaction monitoring-mass spectrometric immunoassay responsive to parathyroid hormone and related variants. *Clinical Chemistry, 56* (2), 281–290.
21. Magrane, M., & Consortium, U. (2011). UniProt knowledgebase: A hub of integrated protein data. *Database* 2011, bar009.
22. Mathivanan, S., Ji, H., Tauro, B. J., Chen, Y.-S, & Simpson, R. J.

(2012). Identifying mutated proteins secreted by colon cancer cell lines using mass spectrometry. *Journal of Proteomics, 76,* 141–149.

23. Montgomery, S. B., Sammeth, M., Gutierrez-Arcelus, M., Lach, R. P., Ingle, C., Nisbett, J., Guigo, R., & Dermitzakis, E. T. (2010). Transcriptome genetics using second generation sequencing in a Caucasian population. *Nature, 464* (7289), 773–777.
24. Nesvizhskii, A. I., Roos, F. F., Grossmann, J., Vogelzang, M., Eddes, J. S., Gruissem, W., Baginsky, S., & Aebersold, R. (2006). Dynamic spectrum quality assessment and iterative computational analysis of shotgun proteomic data: toward more efficient identification of post-translational modifications, sequence polymorphisms, and novel peptides. *Molecular & Cellular Proteomics, 5*(4), 652–670.
25. Nie, S., Yin, H., Tan, Z., Anderson, M. A., Ruffin, M. T., Simeone, D. M., & Lubman, D. M. (2014). Quantitative analysis of single amino acid variant peptides associated with pancreatic cancer in serum by an isobaric labeling quantitative method. *Journal of Proteome Research, 13* (12), 6058–6066.
26. Nørregaard Jensen, O. (2004). Modification-specific proteomics: Characterization of post-translational modifications by mass spectrometry. *Current Opinion in Chemical Biology, 8* (1), 33–41.
27. O'Donovan, C., Apweiler, R., & Bairoch, A. (2001). The human proteomics initiative (HPI). *Trends in Biotechnology, 19* (5), 178–181.
28. Omenn, G. S., Lane, L., Lundberg, E. K., Beavis, R. C., Nesvizhskii, A. I., & Deutsch, E. W. (2015). Metrics for the Human Proteome Project 2015: Progress on the human proteome and guidelines for high-confidence protein identification. *Journal of Proteome Research, 14* (9), 3452–3460.
29. Paik, Y. K., Omenn, G. S., Uhlen, M., Hanash, S., Marko- Varga, G., Aebersold, R., Bairoch, A., Yamamoto, T., Legrain, P., Lee, H. J., Na, K., Jeong, S. K., He, F., Binz, P. A., Nishimura, T., Keown, P., Baker, M. S., Yoo, J. S., Garin, J., Archakov, A., Bergeron, J., Salekdeh, G. H., & Hancock, W. S. (2012). Standard guidelines for the chromosome-centric human proteome project. *Journal of Proteome Research, 11* (4), 2005–2013.
30. Pandey, A., & Pevzner, P. A. (2014). Proteogenomics. *Proteomics, 14* (23–24), 2631–2632.
31. Salisbury, B. A., Pungliya, M., Choi, J. Y., Jiang, R., Sun, X. J., &

Stephens, J. C. (2003). SNP and haplotype variation in the human genome. *Mutation Research, 526*, 53–61.

32. Schandorff, S., Olsen, J. V., Bunkenborg, J., Blagoev, B., Zhang, Y., Andersen, J. S., & Mann, M. (2007). A mass spectrometry-friendly database for cSNP identification. *Nature Methods, 4,* 465–466.
33. Sherry, S. T., Ward, M. H., Kholodov, M., Baker, J., Phan, L., Smigielski, E. M., & Sirotkin, K. (2001). dbSNP: The NCBI database of genetic variation. *Nucleic Acids Research, 29,* 308–311.
34. Shi, T., Fillmore, T. L., Sun, X., Zhao, R., Schepmoes, A. A., Hossain, M., Xie, F., Wu, S., Kim, J. S., Jones, N., Moore, R. J., Pasa-Tolic, L., Kagan, J., Rodland, K. D., Liu, T., Tang, K., Camp, D. G., 2nd, Smith, R. D., & Qian, W. J. (2012). Antibody-free, targeted mass-spectrometric approach for quantification of proteins at low picogram per milliliter levels in human plasma/serum *Proceedings of the National Academy of Sciences of the United States of America, 109* (38), 15395–15400.
35. Sjoblom, T., Jones, S., Wood, L. D., Parsons, D. W., Lin, J., Barber, T. D., et al. (2006). The consensus coding sequences of human breast and colorectal cancers. *Science, 314* (5797), 268–274.
36. Song, C., Wang, F., Cheng, K., Wei, X., Bian, Y., Wang, K., Tan, Y., Wang, H., Ye, M., & Zou, H. (2014). Large-scale quantification of single amino-acid variations by a variation-associated database search strategy. *Journal of Proteome Research, 13* (1), 241–248.
37. Sordella, R., Bell, D. W., Haber, D. A., & Settleman, J. (2004). Gefitinib-sensitizing EGFR mutations in lung cancer activate anti-apoptotic pathways. *Science, 305* (5687), 1163–1167.
38. Su, Z. D., Sun, L., Yu, D. X., Li, R. X., Li, H. X., Yu, Z. J., Sheng, Q. H., Lin, X., Zeng, R., & Wu, J. R. (2011). Quantitative detection of single amino acid polymorphisms by targeted proteomics. *Journal of Molecular Cell Biology, 3* (5), 309–315.
39. Sun, T., Zhou, Y., Yang, M., Hu, Z., Tan, W., Han, X., Shi, Y., Yao, J., Guo, Y., Yu, D., Tian, T., Zhou, X., Shen, H., & Lin, D. (2008). Functional genetic variations in cytotoxic T-lymphocyte antigen 4 and susceptibility to multiple types of cancer. *Cancer Research, 68* (17), 7025–7034.
40. Tabas-Madrid, D., Alves-Cruzeiro, J., Segura, V., Guruceaga, E., Vialas, V., Prieto, G., Garcia, C., Corrales, F. J., Albar, J. P., & Pascual-Montano, A. (2015). Proteogenomics dashboard for the human

proteome project. *Journal of Proteome Research, 14* (9), 3738–3749.

41. Taylor, R. W., & Turnbull, D. M. (2005). Mitochondrial DNA mutations in human disease. *Nature Reviews Genetics, 6* (5), 389–402.
42. Végvári, Á., Sjödin, K., Rezeli, M., Malm, J., Lilja, H., Laurell, T., & Marko-Varga, G. (2013). Identification of a novel proteoform of prostate specific antigen (SNP-L132I) in clinical samples by multiple reaction monitoring. *Molecular & Cellular Proteomics, 12* (10), 2761–2773.
43. Wang, X., & Zhang, B. (2013). CustomProDB: An R package to generate customized protein databases from RNA-Seq data for proteomics search. *Bioinformatics (Oxford, UK), 29* (24), 3235–3237. Wang, Q., Chaerkady, R., Wu, J., Hwang, H. J., Papadopoulos, N., Kopelovich, L., Maitra, A., Matthaei, H., Eshleman, J. R., Hruban, R. H., Kinzler, K. W., Pandey, A., & Vogelstein, B. (2011). Mutant proteins as cancer-specific biomarkers. *Proceedings of the National Academy of Sciences, 108* (6), 2444–2449.
44. Whiteaker, J. R., Zhao, L., Zhang, H. Y., Feng, L. C., Piening, B. D., Anderson, L., & Paulovich, A. G. (2007). Antibody-based enrichment of peptides on magnetic beads for mass-spectrometry-based quantification of serum biomarkers. *Analytical Biochemistry, 362* (1), 44–54.
45. Whiteaker, J. R., Zhao, L., Abbatiello, S. E., Burgess, M., Kuhn, E., Lin, C., Pope, M. E., Razavi, M., Anderson, N. L., Pearson, T. W., Carr, S. A., & Paulovich, A. G. (2011). Evaluation of large scale quantitative proteomic assay development using peptide affinity-based mass spectrometry. *Molecular & Cellular Proteomics, 10* (4), M110.005645.
46. Wood, L. D., Parsons, D. W., Jones, S., Lin, J., Sjoblom, T., Leary, R. J., Shen, D., Boca, S. M., Barber, T., Ptak, J., Silliman, N., Szabo, S., Dezso, Z., Ustyanksky, V., Nikolskaya, T., Nikolsky, Y., Karchin, R., Wilson, P. A., Kaminker, J. S., Zhang, Z., Croshaw, R., Willis, J., Dawson, D., Shipitsin, M., Willson, J. K., Sukumar, S., Polyak, K., Park, B. H., Pethiyagoda, C. L., Pant, P. V., Ballinger, D. G., Sparks, A. B., Hartigan, J., Smith, D. R., Suh, E., Papadopoulos, N., Buckhaults, P., Markowitz, S. D., Parmigiani, G., Kinzle, K. W., Velculescu, V. E., & Vogelstein, B. (2007). The genomic landscapes of human breast and colorectal cancers. *Science (New York, NY), 318* (5853), 1108–1113.
47. Xi, H., Park, J., Ding, G., Lee, Y. H., & Li, Y. (2009). SysPIMP: The web-based systematical platform for identifying human disease-related

mutated sequences from mass spectrometry. *Nucleic Acids Research, 37* (Database issue), D913–920.

48. Yan, H., Yuan, W., Velculescu, V. E., Vogelstein, B., & Kinzler, K. W. (2002). Allelic variation in human gene expression. *Science, 297* (5584), 1143.

CHAPTER 6

PROTEOGENOMICS ANALYSIS IN CANCER RESEARCH

CONTENTS

6.1. INTRODUCTION

Worldwide, cancer is one of the principal grounds of death, accounting for 15% of the total sum of yearly deaths. Additionally, in the following two decades, cancer mortality rates are anticipated to undergo two folds increase. There are more than 200 cancer sorts and each can be categorized into numerous subtypes with diverse molecular and clinical features (Tomczak et al., 2015). This complexity clarifies the heterogeneous reaction to therapy and patients expected survival rate (McDermott et al., 2011). DNA sequence mutations motivate the neoplastic transformation and grounds, among other effects, the uncontrolled cell growth (Hanahan and Weinberg, 2000). Hence, the recognition of the whole catalogue of DNA aberrations becomes the main concern, as it will be the foundation for enhancing not only cancer prevention and its early detection but also its management.

The human genome sequencing (Lander et al., 2001; Venter et al., 2001) was the leading step towards the comprehension of the complexity of human biology. Over current years, the expansion of high-throughput technologies for the description of normal and cancer samples has produced a solid perspective of the human genome and the human proteome (Chin et al., 2011a). For instance, gene expression profiling using DNA microarrays (Cordero et al., 2007) has allowed for the measurement of the expression level of massive number of genes in a single experiment, leading to the existence of genomic signatures that can be used to categorize cancers (Sotiriou and Pusztai, 2009; Desmedt et al., 2009).

Nevertheless, the expansion and utilization of next generation sequencing (NGS) technology have augmented the detection of genomic mutations, such as insertions, deletions, substitutions or amplifications (Meyerson et al., 2010; Trapnell et al., 2013). Whole-genome and whole-exome sequencing have demonstrated to be the most appreciated methods to recognize significant mutations for the diagnosis and treatment of many human disorders, comprising cancer (Pabinger et al., 2014). The recognition of germline mutations (Kurian et al., 2014) and the demonstration of somatic mutations in cancer (Tamborero et al., 2013) are the most shared applications for these experiments.

The advancement in the analysis of cancer genomes has allowed the development of large-scale clarification schemes, comprising the International Cancer Genome Consortium (ICGC, https://dcc.icgc.org), the Cancer Genome Project (CGP, http://www.sanger.ac.uk/genetics/CGP) and The Cancer Genome Atlas (TCGA, http://cancergenome.nih.gov). Despite the datasets produced in these projects are publically obtainable, access to definite specific information is under control as a guard of patient privacy (Chin et al., 2011b). Since it commenced in 2006, the TCGA project has clarified more than 10,000 tumor samples from 34 different cancer sorts, utilizing high-throughput sequencing methods. The bioinformatics data analyses of this big yield of data have allowed for the recognition of more than ten million mutations.

Proteogenomics, as an emerging field that integrates proteomics, genomics and transcriptomics with the aim of better understanding cellular functions, (Faulkner et al., 2015; Nagaraj et al., 2011) allowed for the study of the proteome of non-model species, where the availability of a reference protein database is limited (Evans et al., 2012). Interestingly, the major advances in proteogenomics have been made in cancer research, and more specifically in the detection of tumor-specific changes in the proteome.

Changes such as single amino acid polymorphisms (SAPs) may lead to disease initiation, progression or discrepancy in reaction to treatment (Alfaro et al., 2014; Zhang et al., 2014). The serious stage in the cancer samples analysis is the generation of the customized peptide databases needed to execute the MS searches. Diverse approaches grounding on the processing of genomic data, usually utilizing DNA-Seq or RNA-Seq experiments, have been illustrated (Woo et al., 2014; Nesvizhskii, 2014; Wang and Zhang, 2013). In this setting, the big number of experiments accessible in projects such as the TCGA is a predominantly powerful resource to fuse this methodology in the clinical oncology background.

The Human Proteome Project (HPP) initiative (Legrain et al., 2011) has excuted a leading role in the incorporation of transcriptomics and proteomics data (Paik and Hancock, 2012; Segura et al., 2014) and the expansion of new proteogenomics tools and bioinformatics methods is presently a zone of active research in the consortium (Krasnov et al., 2015; Nagaraj et al., 2015; Tabas-Madrid et al., 2015).

6.2. BIOINFORMATICS TOOLS FOR SAP IN CANCER DETECTION

In this part, we recommend a workflow of bioinformatics to produce proteomic databases for the target of recognizing SAPs in cancer samples grounding on the knowledge of somatic mutations drived from the TCGA project.

6.2.1. The Cancer Genome Atlas

The Cancer Genome Atlas (TCGA) is an integrative project between the National Human Genome Research Institute (NHGRI) and the National Cancer Institute (NCI). The main target of the project is to demonstrate the main genomic alterations in various sorts of cancers with the purpose of understanding the molecular foundation of the disease (Tomczak et al., 2015). To execute this target, high throughput technologies have been implemented, comprising next generation sequencing (NGS) and microarrays. Since it's beginning in 2006, the TCGA has investigated normal tissue and matched tumors from 11,000 patients to analyze 34 cancer sorts and subtypes. The accessibility of these cancer genomic datasets is predictable to enhance the diagnosis, prevention, and treatment of cancer.

The structure of TCGA comprises numerous centers accountable for the collection of sample and processing, high-throughput sequencing and bioinformatic tools for analysis of the accessible datasets. As shown in Table 6.1, 10,183 exomes were included from 31 cancer sorts for the production of the SAP database. Three cancer type samples were removed: MESO (there are no available exome data), LAML (hg18 genome was the base of the results) and STAD (the results are not delivered utilizing the standard MAF file). The data of somatic mutation (MAF files) were derived from the data portal of the TCGA project and processed utilizing in-house R scripts. Without any filtering process, 3,009,480 unique variants were recognized in the dataset. The cancer type with the highest number of detected variants was LIHC (with 956,761 mutations) while the cancer type with the fewest number of detected variants was UVM (with 3,918 mutations). The mean number of variants per analyzed exome was 296 while 97,080 mutations were the mean number of mutations per cancer type.

Table 6.1. Cancer Types Studied in the TCGA Project

Cancer type	**Acronym**	**#cases/#exomes**	**#mutations**
Acute Myeloid Leukemia	LANIL	200/150	-
Adrenocortical carcinoma	*ACC*	80/80	18,052
Bladder Urothelial Carcinoma	BLCA	412/412	158,417
Brain Lower Grade Glioma	LGG	516/516	59,419
Breast invasive carcinoma	BRCA	1,098/1,081	94,008
Cervical squamous cell carcinoma and endocervical adenocarcinoma	CESC	308/305	82,680

Cholangiocarcinoma	CHOL	36/36	7,679
Colon adenocarcinoma	COAD	461/458	117,118
Esophageal carcinoma	ESCA	185/184	81,248
FFPE Pilot Phase II	FPPP	38/38	40,110
Glioblastoma multiforme	GBM	528/512	60,392
Head and Neck squamous cell carcinoma	HNSC	528/510	137,454
Kidney Chromophobe	KICH	66’66	8,391
Kidney renal clear cell carci-noma	KIRC	536/520	42,196
Kidney renal papillary cell carcinoma	KIRP	291/288	30,320
Liver hepatocellular carcinoma	LIHC	377/375	956,761
Lung adenocarcinoma	LUAD	521/517	219,852
Lung squamous cell carcinoma	LUSC	504/497	65,065
Lymphoid Neoplasm Diffuse Large B-cell Lymphoma	DLBC	48/48	16,462
Mesothe I ioma	MESO	87/0	0
Ovarian serous cystadenocarci-noma	OV	586/536	12,681
Pancreatic adenocarcinoma	PAAD	185/184	56,871
Pheochromocytoma and Para-ganglioma	PCPG	179/179	6,818
Prostate adenocarcinoma	PRAD	498/498	24,399
Rectum adenocarcinoma	READ	171/168	32,815
Sarcoma	SARC	2611255	77,698
Skin Cutaneous Melanoma	SKCM	470;470	298,752
Stomach adenocarcinoma	STAD	443/441	
Testicular Germ Cell Tumors	TGCT	150/150	25,077
Thymoma	THYM	124/123	37,640
Thyroid carcinoma	TIICA	507/496	20,684
Uterine Carcinosarcoma	UCS	57/57	13,048
Uterine Corpus Endometrial Carcinoma	UCEC	548/544	203,455
Uveal Melanoma	UV NI	80/80	3,918

The SNV calls collected from the TCGA were further processed and integrated into cancer mutations database. The detected variants were re-annotated using VeP (McLaren et al., 2010) to confirm annotation consistency across all exome findings, to predict the effect of the mutations. This software converts SNVs to the related amino acid coordinates and yields the SIFT (Kumar et al., 2009) and PolyPhen-2 (Adzhubei et al., 2010) scores for each SNV. This information was used to filter the synonymous SNVs utilizing the R/Bioconductor statistical analysis. We only retained only those SNVs classified as "missense" were retained. After the filtering process, 1,161,751 distinctive genomic variants distributed across the cancer sorts were accessed as shown in Figure 5.6a.

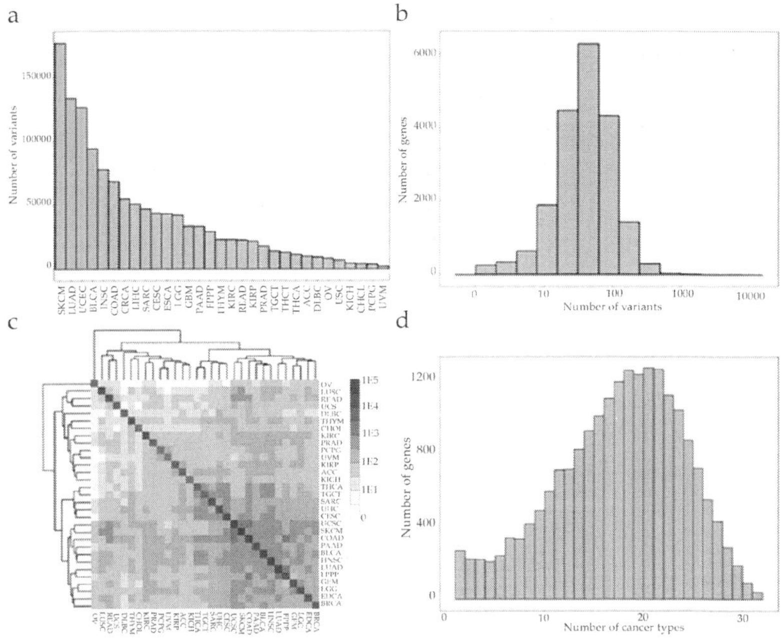

Figure 6.1. Analysis of the features of the non-synonymous variants recognized in the TCGA dataset annotated using VeP.

CHOL, PCPG, and UVM were the cancer types with the fewest number of identified mutations while SKCM, LUAD, and UCEC were the cancer types with the highest. The association between the count of genes and the number of mutations detected in every single gene is illustrated in Figure 6.1b. Notably, more than 12,000 genes had at least 50 variants. In further studies, the number of frequent variants between different tumor sorts was analyzed in order to launch a classification of cancers according to their mutations. This analysis has shown that the clustering of the matrix of the number of mutations shared between each pair of cancer sorts (Figure 6.1c) clearly identify tumor specific mutation genes (256 genes) and ubiquitous mutant genes, demarcated as those genes that are mutated in all the studied cancer types (37 genes).

Finally, in the same study, the explanation of the directory of somatic mutations considered for proteogenomics analysis was executed by defining the percentage of samples in which each gene is changed (Figure 6.2a). Remarkably, it was found that most of the genes are mutated in a very low percentage of the dataset. Exactly, only 43 genes were mutated in more than 5% of the TCGA cancer samples (Figure 6.2b). This group comprised genes such as TTN (titin), TP53 (tumor protein p53), BRAF (v-raf murine sarcoma viral oncogene homolog B1), PIK3CA (phosphoinositide-3-kinase, catalytic, alpha polypeptide) and KRAS (v-Ki-ras2 Kirsten rat sarcoma viral oncogene homolog); all of these genes have been demarcated as commonly mutated genes in preceding analyses of the TCGA database (Ciriello et al., 2013; Kandoth et al., 2013). The intense analysis of this group of genes utilizing Ingenuity software (www.ingenuity.com) permitted the identification of enriched functional clusters such as Cancer, Cell Death, survival and cell cycle, DNA Repair or Cellular reaction to therapies, which affirmed the effect of these genes in cancer-related pathways.

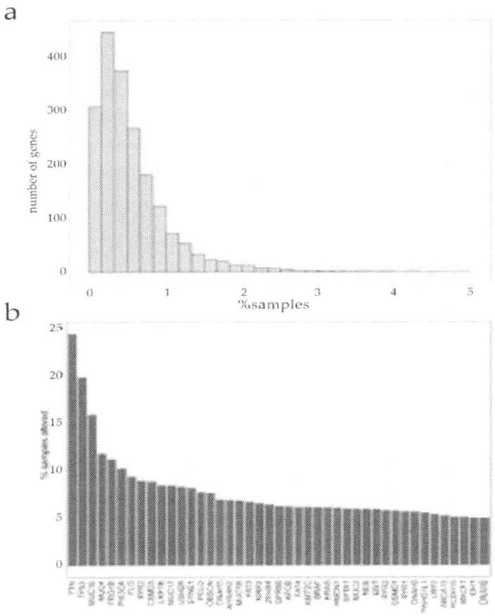

Figure 6.2 (a) genes distribution as a function of the percentage of the TCGA samples in which they are mutated. (b) Most commonly mutated genes found comprising well-known oncogenes and tumor suppressors.

6.2.2. SAP Database Generation

The mutated peptides of the SAP database were produced from the non-synonymous variants of the TCGA samples utilizing the following method (Yang and Lazar, 2014). Primarily, sequences of 80 amino acids flanking the mutated amino acid were extrapolated from the FASTA file of protein sequences (Ensembl version 75). The original amino acid and the position of the mutation were verified, filtering out cases in which an inconsistency was demonstrated. The mutated peptides only stored in the database were the mutated peptides whose sequences did not exist in the reference proteome. The dasHPPboard (Tabas-Madrid et al., 2015), a bioinformatics tool that provides access to resources for proteogenomics analyses based on ENCODE (ENCODE Project Consortium et al., 2011) and Illumina Human Body Map (HBM) datasets, was implemented successfully to provide a similar workflow.

In this way, a total of 1,525,055 unique SAPs located in 92,412 proteins (corresponding to 19,925 genes) were identified with a mean number of SAPs per gene was 77. UVM was the cancer type with a minimum number of SAPs (2,738 mutated peptides) while SKCM was the cancer type with the maximum number of SAPs (222,665 mutated peptides). Before the creation of the FASTA file required to execute the protein searches, the mutation redundancy from the database was removed. In addition, to identify the non-redundant peptides of this database, numerical identifiers were used giving the option of customizing the content of the FASTA file produced. Finally, the results were clearly annotated with in house programmed R scripts.

6.2.3. Annotation of Genetic Variants

Once the database of mutations was completed, a clinical and bibliographic annotation of the genetic variants was integrated into the bioinformatic workflow. This information was utilized to complement the yield of VeP and stratify the priority of the detected variants. In order to simplify the interpretation of the results two databases of reliable genetic data were designated and processed: ClinVar (Landrum et al., 2014) and COSMIC (Forbes et al., 2015). COSMIC (Catalog Of Somatic Mutations In Cancer, http://cancer.sanger.ac.uk) is the complete store of somatic mutations. The release comprised in the analysis pipeline (v74; September 2015) illustrated 3,480,051 coding mutations in about a million tumor samples and through most of the human genes. ClinVar (http://www.ncbi.nlm.nih.gov/clinvar) is another database of information on the associations among genetic variants and phenotypes meticulously associated with dbVar (Lappalainen et al., 2013) and dbSNP (Smigielski et al., 2000). The integrated release (November 2015) comprised 134,321 variations in 26,406 genes.

The study was able to annotate 57.59% of the nonsynonymous variants of the TCGA, 0.39% using ClinVar and 57.47% using COSMIC (Figure 6.3a). Putting the focus on the ClinVar annotations, it was demonstrated that 25.36% of the variants previously illustrated in this database were pathogenic or probable pathogenic, while 28.76% were benign or probable benign (Figure 6.3b).

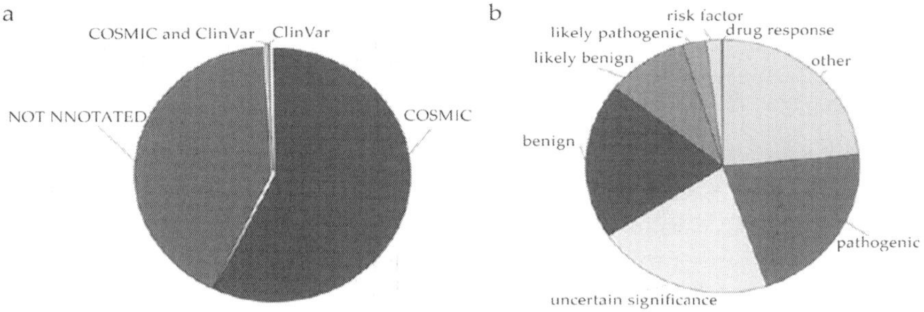

Figure 6.3 (a) Annotation of the mutated peptides integrated to the SAP database. (b) Classification of the mutated peptides annotated using ClinVar

6.3. PROTEOGENOMICS METHODS FOR THE IDENTIFICATION OF SAPS USING SHOTGUN EXPERIMENTS

Based on the FASTA databases created from the TCGA datasets, the analysis of high throughput proteomic experiments to identify mutated peptides in cancer samples was executed. This part is given to the proteogenomics study of special four cancer cell lines employing public shotgun experiments. Different combinations of search engines and SAP databases were used to further check that the proteogenomics approach illustrated is applicable and is able to give consistent and coherent results.

6.3.1. Public Shotgun Experiments

Four human cell line experiments available in the PRIDE database were analyzed (Vizcaíno et al., 2013). This experimental work was provided by the Spanish HPP consortium to the ProteomeXchange repository (http://www.

proteomexchange.org/) with accession numbers PXD000039, PXD000442, PXD000443, and PXD000449. Here, 2 replicates were selected for each cell line: CCD18 (human colon fibroblast cell line), MCF7 (human breast adenocarcinoma cell line), Jurkat (human T cell lymphoblast-like cell line) and Huh7 (human hepatocellular carcinoma cell line). Further analysis was done through the use of Mascot generic files (mgf).

6.3.2. Shotgun Data Analysis

Two different proteomic analyses were executed for each sample: the first is utilizing the protein reference database (ProteinDB), which is the Ensembl protein database, while the other employing the previously produced database of mutated peptides (SAPDB) as illustrated in Figure 6.4. In the former search, the proteins present in the samples were identified through the application of a FDR criterion at the PSM and protein level. A new search was executed after removal of the assigned spectra from the mgf files to identify mutated peptides employing an SAP database. In this step, the statistical analyses of the findings were performed at PSM level only.

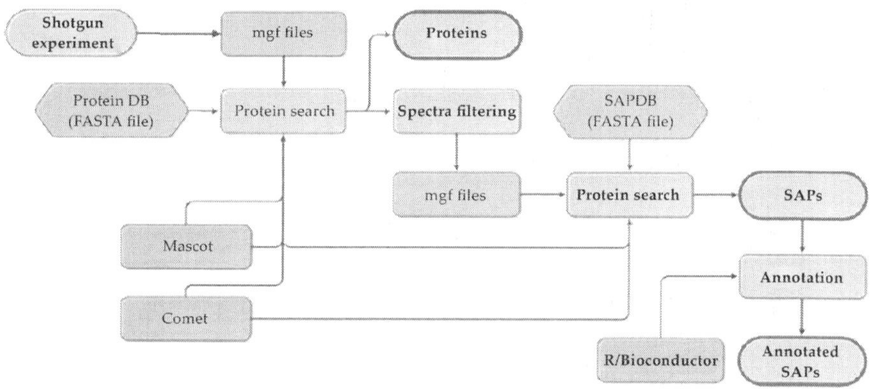

Figure 6.4. Bioinformatics workflow analysis for SAP recognition using proteomic shotgun experiments.

6.3.2.1. Protein Identification

All the mgf files downloaded from PRIDE were searched in relation to the ProteinDB database (Ensembl version 75) employing the target-decoy strategy with Comet (Eng et al., 2013) and an in-house Mascot Server v. 2.3 (Matrix Science, London, U.K.) search engines. A decoy database was generated utilizing the peptide pseudo-reversed approach and separate searches were executed for target and decoy databases. Search parameters were defined as follows: oxidized methionine as variable modification and carbamidomethyl cysteine as a fixed modification. Precursor and fragment mass tolerance were put to 10 ppm and 0.05 Da correspondingly, and 2 missed cleavages were permitted. Using Mayu (Reiter et al., 2009), False Discovery Rates at the PSM and protein levels were calculated.

Protein identifications were accessed through the application of the criteria of PSM FDR < 1% and protein FDR < 1% sticking to the C-HPP guidelines. Using the PAnalyzer algorithm (Prieto et al., 2012) Protein inference for Mascot results was executed, and non-conclusive protein groups were discarded. Regarding Comet, the protein accessions were provided by the search engine. Table 6.1 summarizes all the results received in the analysis of the proteome of the cell lines undergone study analysis. The ratio of assigned spectra was quite low in all the cell lines (below 15% of the received spectra). The count of proteins and genes detected with Comet was remarkably lower than the number perceived with Mascot, possibly due to the dissimilar protein inference algorithm used in both cases.

Table 6.2. Count of Peptides, Proteins, and Genes Detected Utilizing ProteinDB with Mascot and Comet Search Engines (Eng et al., 2013)

	HUH7_1	HUH7_2	MCF7_1	MCF27_2	JURKAT_1	JURKAT_2	OCD18_1	CCD18_2
Spectra	589,694	779,849	452,970	468,831	727,488	727,488	831,689	1,492,124
Mascot								
PSMs	27,525	52,389	461,492	130,989	527,701	435,383	538,888	543,506
Peptides	1,367	3,303	28,573	12,175	39,433	36,026	17,981	17,152
Proteins	3,596	6,159	14,000	11,324	19,094	17,888	11,581	11,005
Genes	1,017	1,890	4,785	3,669	6,409	6,110	3,837	3,771
Assigned spectra	4,501	8,444	75,405	22,508	94,976	77,885	80,078	83,025
Comet								
PSMs	2,531	8,332	66,767	2,365	99,214	43,981	72,087	38,349
Peptides	759	3,176	25,418	1,227	39,142	20,922	15,836	9,842
Proteins	541	1,473	4,212	560	6,218	4,413	2,951	3,170
Genes	524	1,394	3,722	548	5,357	3,990	2,680	2,897
Assigned spectra	2,518	7,482	65,550	2,349	97,487	43,401	69,987	37,701

Despite the results may not be adequately comparable, most of the proteins recognized in this study were recognized using both search engines.

In order to enhance the evaluation and the assessment of the results, the protein accession codes from Ensembl (ENSP) were transformed into gene accessions (ENSG). The analysis with Mascot yielded 8,063 different genes, 21.92% of which are common to the 4 cell lines and 33.51% were specific to only one (Figure 6.5a). In parallel, the analysis with Comet yielded 6,632 genes, 17.55% were common genes and 39.32% were specific genes (Figure 6.5b). The comparison between the sets of genes perceived in each sample with Mascot and Comet demonstrated that the count of common identifiers is more for those samples showing the same biological source (Figure 6.5c). Putting into consideration all the protein identifications, the ratio of genes detected with both search engines was as high as 69.51% even though the protein groups were calculated using dissimilar approaches (Figure 6.5d).

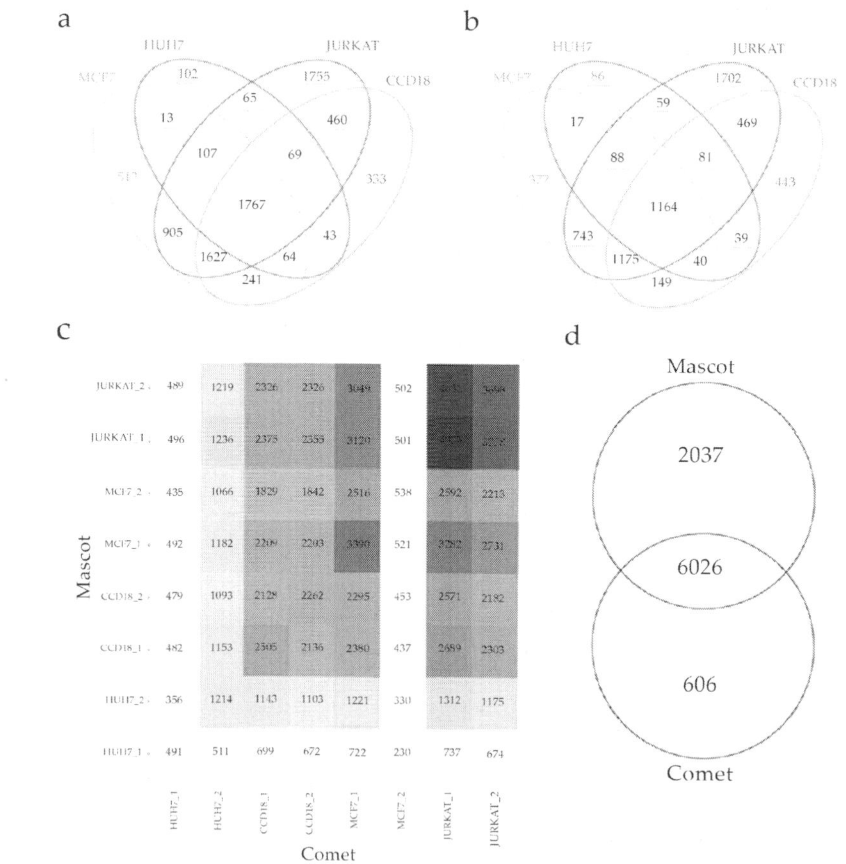

Figure 6.5. ProteinDB search findings of the shotgun experiments in the MCF7, CCD18, Jurkat, and Huh7 cell lines at the protein level.

6.3.2.2. *SAP Detection*

The mutated peptides detection began with the removal of the spectra assigned to proteins from the unique mgf files. The new datasets were analyzed employing the complete SAP database (TCGA) which comprised 1,525,055 unique SAPs and cancer-specific databases. Therefore, four extra databases were generated: a COAD database with 89,771 unique SAPs for the analysis of CCD18, a LICH database with 67,694 unique SAPs for the analysis of Huh7 and a BRCA database with 72,249 unique SAPs for the analysis of MCF7. Reagrding the Jurkat cell line only the complete TCGA database was employed due to the shortage of a unique cancer dataset. Filtered mgf files were searched using Comet and Mascot with the same query factors employed for the recognition of proteins. However, only PSM FDR < 1% was applied here as a statistical threshold for recognizing mutated peptides. The results received in each of the analysis performed are summarized in Table 6.3.

Table 6.3. Number of Mutated Peptides, Proteins, and Genes Identified Using SAPDB

Search engine	SAP Database	Sample	Assigned spectra	PSMs	SAPS	Mutated proteins	Mutated genes
Comet	BRCA	MCF7_1	83	84	66	185	65
		MCF7_2	136	137	122	352	115
	COAD	CCD18_1	44	52	14	63	14
		CCD18_2	467	475	223	571	179
	LBW	IFIUFf7_1	1	1	1	4	1
		HUH7_2	446	446	260	704	235
	TCGA	CCD18_1	909	941	241	578	200
		CCD18_2	2,649	2,781	1,051	1,474	481
		HUH7_1	37	38	18	44	18
		HUH7_2	553	567	318	770	242
		JURKAT 1	407	426	234	552	193
		JURKAT 2	121	123	82	216	78
		MCF7_1	426	435	183	454	151
		MCF7_2	5	5	5	14	5

Mascot	BRCA	MCF7_1	81	115	97	218	68
		MCF7_2	18	22	8	24	5
	COAD	CCD18_1	35	65	19	64	10
		CCD18_2	52	82	33	78	21
	LIHC	HUH7_1	4	5	5	9	4
		HUH7_2	6	6	5	9	5
	TCGA	CCD18_1	596	907	316	581	179
		CCDI8_2	893	1,238	466	841	256
		HUH7_1	90	134	50	107	35
		HUH7_2	117	153	105	225	70
		JURKAT 1	656	1,111	690	1,150	332
		JURKAT 2	398	554	364	795	239
		MCF7_1	565	1,972	1,627	1,093	343
		MCF7_2	16	27	22	63	12

In spite of the large count of spectra reserved upon filtering, the results realized in terms of PSM and recognized features (mutated peptides, proteins and genes) were very low. These results were coherent with the expected scarcity of mutated peptides in the samples and the random way of MS experiments. For each selection of search engine (Mascot or Comet) and SAP database (TCGA or tumor specific), the results were compared (genes or mutated peptide) in each cell line (Figure 6.6). In almost all cases, the count of detected characteristics was higher when the search was executed with Comet. In addition, it is imperative to emphasize the higher number of recognitions yielded using the complete SAPs directory than when employing cancer-unique databases. What is even more noticeable is the reduced count of mutated genes in comparison to the related number of SAPs, although this is probably due to the count of genetic variants per gene.

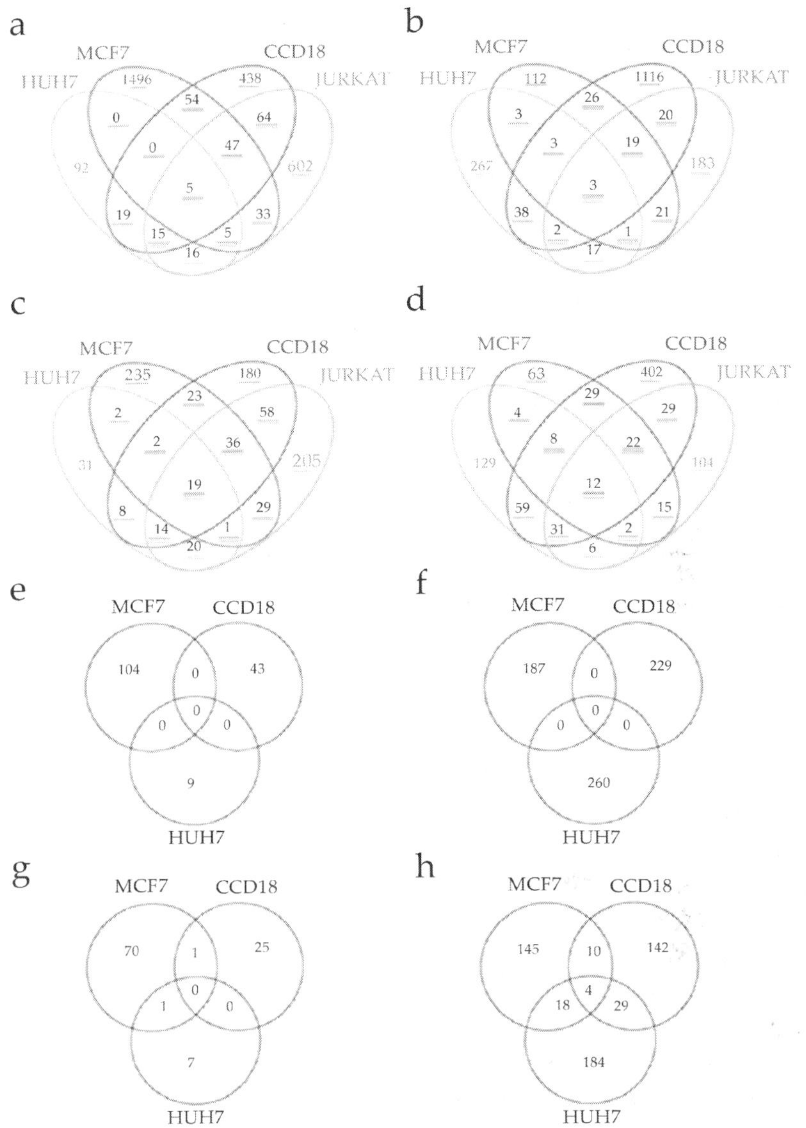

Figure 6.6. Venn diagrams demonstrating the findings of the identification of SAPs and mutated genes using shotgun proteomic data of MCF7, Jurkat, CCD18 and Huh7 cell lines.

A remarkable point here is the degree of resemblance between the search results attained using the complete TCGA database and the tumor-unique databases (Figure 6.7). For a given cell line a large number of the mutated peptides was recognized using both databases. Yet, some of the SAPs were common with another cell line but they were not perceived using both databases. This may be related to the presence of the same genetic variants in different cancer types and the influence of the size of the database on the FDR calculation.

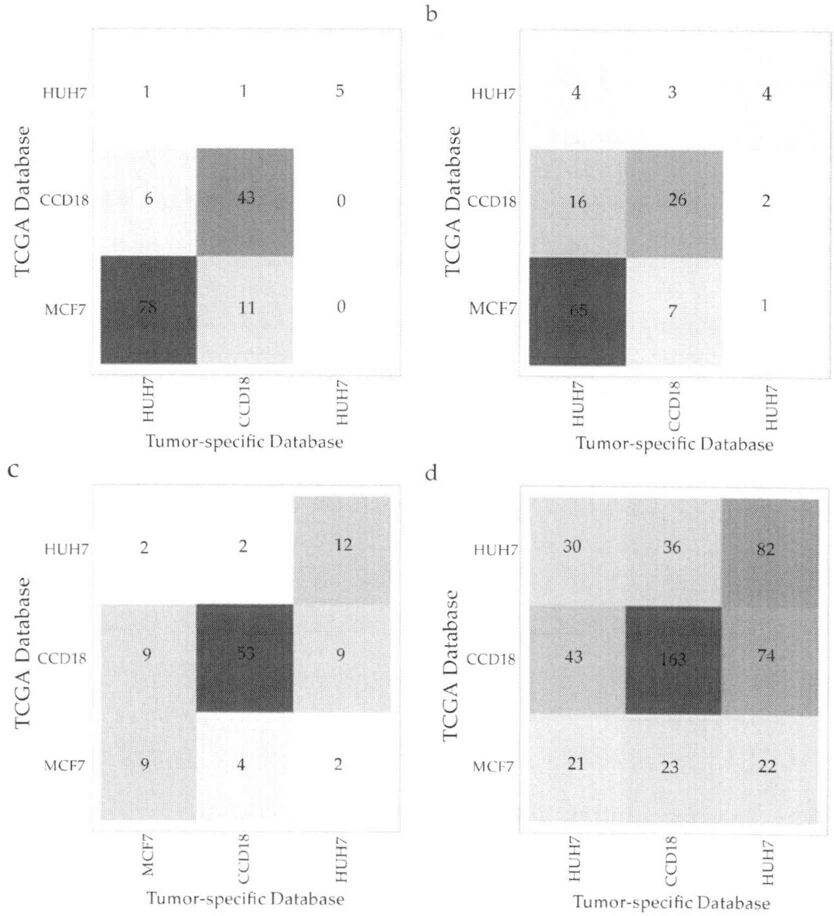

Figure 6.7. Heatmap demonstrating the similarity among the landscape of mutations received using a complete TCGA database or tumor-specific databases.

In a similar way, on comparing the results between search engines taking all the identifications as a sum (Figure 6.8). The ratio of shared

SAPs employing the TCGA database was 11.7% while the ratio of shared SAPs using tumor-unique databases was 3.48%. When the comparison was achieved using the mutated genes perceived, the overlap augmented both when the TCGA database was employed (32.59%) and also when the tumor-specific databases were utilized (6.25%). Yet, the count of matches in tumor-unique questions was inferior to the count of matches in the complete TCGA database. This result proposes that the value of the peptide recognitions received from the complete TCGA database was superior. For the sake of confirmation, the distributions of the search engine scores were represented (XCorr for Comet and ion score for Mascot). As shown in Figure 38e and 38f, the distributions of scores for the searches executed with the TCGA database are clearly superior to the distributions of scores obtained from tumor unique queries.

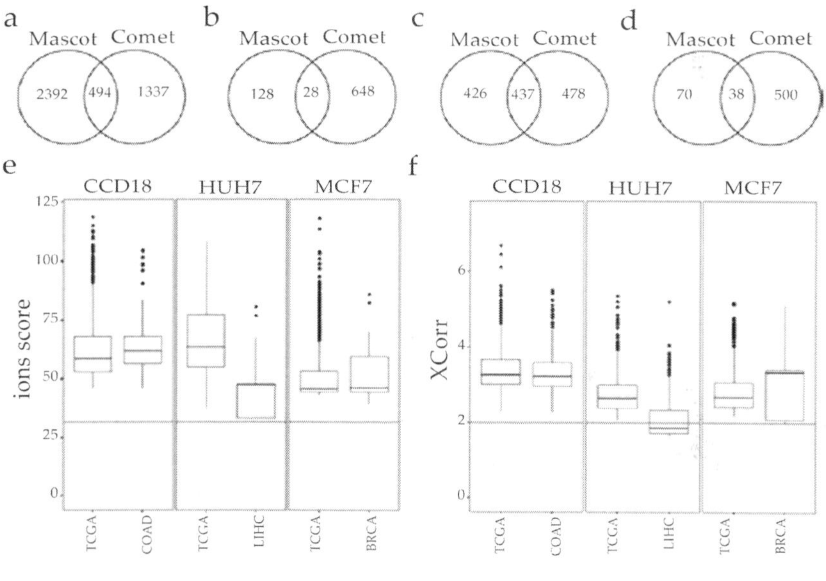

Figure 6.8. Comparison of the SAP detection analysis as a function of the search engine.

In sum, 4,826 mutated peptides in the four cell lines utilizing duplicated shotgun experiments were identified, 2,916 with Mascot and 2,419 with Comet. This can be interpreted as the recognition of 1,580 mutated genes, 1,161 genes detected with Comet and 877 genes detected with Mascot. The study able to annotate 2,422 recognized genetic variants utilizing COSMIC (58.04% of all detections), while only 30 perceived genetic variants (0.69%) were previously illustrated in ClinVar.

6.3.2.3. Functional Analysis of Mutated Genes

The list of mutated genes received after the recognition of SAPs was further analyzed in order to validate their effect in cancer. Using Ingenuity, enrichment analysis of disease categories for the 1,580 mutated genes detected was performed. Of this gene list, 1,252 were linked to the cancer category with a p-value < 1e-30. In addition, among the enriched diseases with at least 300 annotated genes it was found "colorectal cancer" (555 genes), "breast or colorectal cancer" (663 genes), "liver tumor" (523 genes), "hepatocellular carcinoma" (502 genes), and "hematological neoplasia" (353 genes) or very closely related categories (Figure 6.3a). This is a critical result, which demonstrates that the recognized variants are actually connected to the cancer types studied through the public shotgun experiments of cell lines (COAD, BRCA, LALM, and LIHC). A further selection of the mutated genes detected in at least 7 of the different analyses was executed to create a heatmap and clustering represented in Figure 6.3b. The results showed that the count of genes was higher when the complete TCGA database was utilized and the groupings were linked to the cell line analyzed.

6.3.2.4. Curated Analysis of Mutated Peptides

The analysis of the mutated peptides detected was completed with a manual curation of the best PSM potential players. 35 PSMs obtained using Mascot was selected (the best 5 PSMs for each result shown in Figure 6.9e). The related spectra were appraised in a blind manner by two separate MS experts who graded their quality as high, medium or low depending on three features: peak intensities, the quality of y-ion or b-ion series and the presence of no assigned peaks (Jumeau et al., 2015).

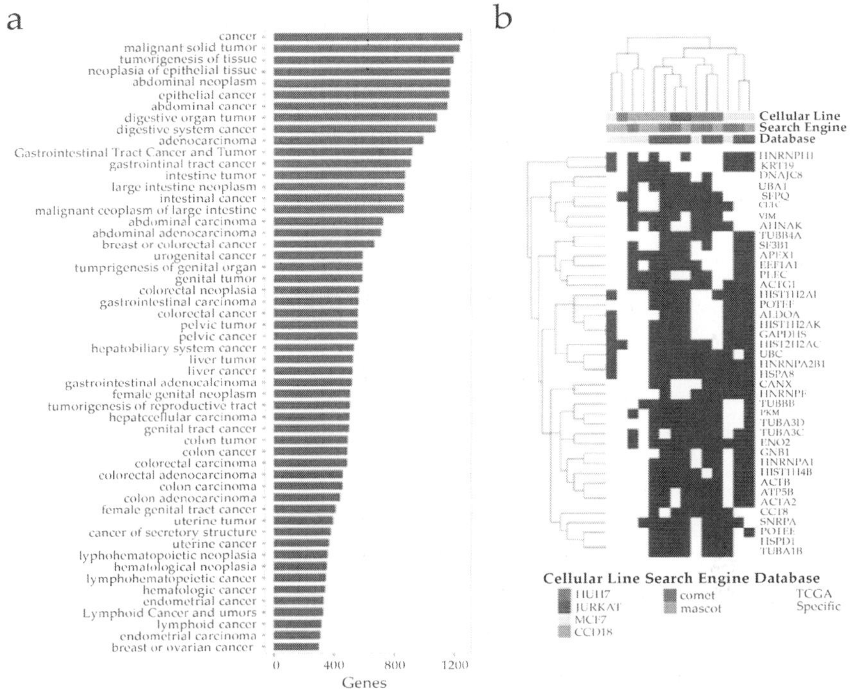

Figure 6.9. (a) Enrichment of disease categories using Ingenuity for the list of 1580 mutated genes detected in the proteogenomics analysis of SAPs. (b) Clustering of the mutated genes detected in at least 7 of the analyses executed.

The outcome of this confirmation process is summarized In Table 6.4, where 65.71% of the spectra were appraised as "medium" or "high" by the mass spectrometry experts, while the 34.29% of the spectra were reflected of "low" quality. This proposes that a significant ratio of the detected matches could be taken into account for further confirmation steps using other approaches, for instance, a SRM/MRM assay.

Two of the mutated peptides detected from Table 6.4 (gene APEX1 in COAD cancer type and SERPINB11 gene in LIHC cancer type) was chosen as an example to examine and assign the fragment ions utilizing the spectrum raw signal (Figure 6.10). In both cases, the y-ion or b-ion series allowed to create the sequence of the peptide accessed with the search engine. It is imperative to highlight that both sequences comprised the mutated amino acid, which confirmed the existence of the SAPs in CCD18 and Huh7 cell lines respectively.

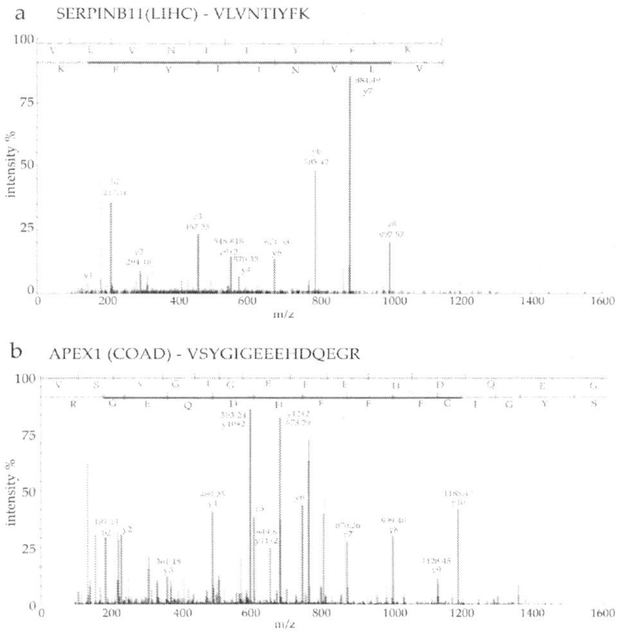

Figure 6.10. (a) Raw spectrum and assigned b-ions and y-ions for the mutated peptide VLVNTIYFK detected in LIHC tumor type in the gene SERPINB11. (b) Raw spectrum and assigned b-ions and y-ions for the mutated peptide VSYGIGEEEHDQEGR detected in COAD tumor type in the gene APEX1 (Table 6.4).

Table 6.4. List of the 35 PSMs Selected for Manual Curation

Gene ID	**Gene name**	**Chr**	**Position**	**WA**	**Cancer type**	**Spectrum quality**
ENSG00000156508	EEFI A I	6	74,228,666	WA	HNSC	Low
ENSG00000184260	HIST2H2AC	I	149,858,837	Q/E	LUAD	Low
ENSG00000147140	NONO	X	70,514,267	VI”	LIHC	Medium
ENSG00000178209	PLEC	8	145,007,187	AN	COAD, CESC	Low
ENSG00000180543	TSPYL5	8	98,289,714	775	COAD	Medium
ENSG00000168090	COPS6	7	99,688,715	R/C	ACC	Medium
ENSG00000144381	HSPDI	2	198,363,449	Q/K	LUAD	High
ENSG00000197157	SNDI	7	127,724,820	A/S	WAD, GBM	High
ENSG00000122566	HNRNPA2BI	7	26,237,265	W/G	LIHC	Low

ENSG00000101558	VAPA	18	9,950,451	M/I	SARC	Medium
ENSG00000144381	HSPDI	2	198363449	Q/K	LUAD	Low
ENSG00000077312	SNRPA	19	41265363	D/Y	UCEC	High
ENSG00000171858	RPS21	20	60,962,906	K/R	CHOL	Low
ENSG00000144381	HSPDI	2	198,363,449	Q/K	LUAD	Low
ENSG00000077312	SNRPA	19	41,265,363	D/Y	UCEC	High
ENSG00000142168	SODI	21	33,040,872	V/A	UCEC	Medium
ENSG00000104833	TUBB4A	19	6,502,196	*GIS*	COAD	High
ENSG00000104833	TUBB4A	19	6,501,408	G/V	WAD	High
ENSG0000017 1314	PGAMI	10	99,190,370	WI.	UCS	High
ENSG00000184640	SEPT9	17	75,488,782	VIC	KIRP	Medium
ENSG00000169045	HNRNPHI	5	179,043,881	G/D	BRCA	Medium
ENSG00000105679	GAPDHS	19	36,034,721	D/Y	BRCA	High
ENSG00000169045	HNRNPHI	5	179,043,881	G/13	BRCA	Low
ENSG00000141837	CACNA IA	19	13,356,070	R/C	BRCA	Medium
ENSG00000122566	HNRNPA2B1	7	26,236,568	E'Q	BRCA	Medium
ENSG00000100823	APEX!	14	20,925,154	ONE	COAD	Medium
ENSG00000088682	COQ9	16	57,486,729	*GIS*	COAD	Low
ENSG0000018 I 873	IBA57	I	228,362,682	*GIS*	COAD	Medium
ENSG00000100823	APEX I	14	20,925,154	INE	COAD	High
ENSG00000178209	NEC	8	145,001,031	WR	COAD	Medium
ENSG00000116560	SFPQ	I	35,656,352	K/T	LIHC	Low
ENSG00000116560	SFPQ	I	35,656,352	Kir	LIHC	Low
ENSG00000206072	SERPINBII	18	61,387,312	MT	LIHC	High
ENSG00000184260	HIST2H2AC	1	149,858,769	R/P	LIHC	Low
ENSG00000162396	PARS2	1	55,223,846	0/13	LIHC	Medium

6.4. SUMMARY

In this section, the detection of mutated peptides in cancer samples using different computational resources, databases and bioinformatic analysis workflow was described. The analysis workflow was separated into two successive parts. First, the database of SAPs was produced utilizing the information in relation to the variants of interest. After the FASTA files needed for peptide recognitions were produced, the remaining part of the analysis was targeting the statistical analysis and the annotation of the findings of the proteomic search work. The database was generated from the data acquired from the TCGA project. In this project, more than 10,000 exomes from more than 30 cancer sorts were analyzed to yield an index of more than three million somatic variants. In order to retain only missense mutations in the SAP database, the VeP software was utilized to infer the effect of the variants. Additionally, the variants were also annotated with clinical significance and bibliographic information employing two reliable resources of somatic mutations in cancer, the ClinVar and COSMIC databases.

The final findings of the data processing were a set of mutations that can be utilized to create on-demand FASTA files. The assortment of mutations could be grounded by tumor type, clinical significance, or another approach of interest. Here, the mutation landscape of 4 cancer cell lines (Huh7, MCF7, CCD18, and Jurkat) was studied to confirm the performance of the suggested approach. Publicly accessible proteomic shotgun experiments were downloaded from PRIDE database and further analyzed employing a reference proteome (Ensembl version 75) and two SAP databases: all TCGA mutated peptides and tumor-unique mutated peptides. In addition, the efficiency of the detection was compared using two frequently used search engines, Mascot and Comet. In the former search for each sample proteins present in the cell lines were identified, using a target-decoy strategy and a FDR threshold of 1% at the PSM and protein levels. Protein inference was executed through the application of the PAnalyzer algorithm in Mascot analyses and the output of the search engine in the Comet analyses.

Despite the results were not completely comparable, substantial overlap existed. Finally, after elimination of the previously assigned spectra a second search was accomplished against the SAP database (complete TCGA data- set or cancer-specific mutations). The list of mutated genes recognized with these analyses was functionally annotated and their effect in cancer, and precisely in the cancer types under study, was confirmed. Another authentication step was performed using two independent experts to appraise the best PSMs detected. This study established the rationality of the proteogenomics method to perceive mutated peptides in cancer samples.

REFERENCES

1. Adzhubei, I. A., Schmidt, S., Peshkin, L., et al. (2010). A method and server for predicting damaging missense mutations. *Nature Methods, 7* (4), 248–249.
2. Alfaro, J. A., Sinha, A., Kislinger, T., et al. (2014). Oncoproteogenomics: Cancer proteomics joins forces with genomics. *Nature Methods, 11* (11), 1107–1113.
3. Ansong, C., Purvine, S. O., Adkins, J. N., et al. (2008). Proteogenomics: Needs and roles to be filled by proteomics in genome annotation. *Briefings in Functional Genomics & Proteomics, 7* (1), 50–62.
4. Chin, L., Andersen, J. N., & Futreal, P. A. (2011a). Cancer genomics: From discovery science to personalized medicine. *Nature Medicine, 17* (3), 297–303.
5. Chin, L., Hahn, W. C., Getz, G., et al. (2011b). Making sense of cancer genomic data. *Genes & Development, 25* (6), 534–555.
6. Ciriello, G., Miller, M. L., Aksoy, B. A., et al. (2013). Emerging landscape of oncogenic signatures across human cancers. *Nature Genetics, 45* (10), 1127–1133.
7. Cordero, F., Botta, M., & Calogero, R. A. (2007). Microarray data analysis and mining approaches. *Briefings in Functional Genomics & Proteomics, 6* (4), 265–281.
8. Desmedt, C., Sotiriou, C., & Piccart-Gebhart, M. J. (2009). Development and validation of gene expression profile signatures in early-stage breast cancer. Cancer Investigation, 27 (1), 1–10.
9. ENCODE Project Consortium, Bernstein, B. E., Birney, E., et al. (2011). A user's guide to the encyclopedia of DNA elements (ENCODE). *PLoS Biology, 9* (4),e1001046.
10. Eng, J. K., Jahan, T. A., & Hoopmann, M. R. (2013). Comet: An open-source MS/MS sequence database search tool. *Proteomics, 13* (1), 22–24.
11. Evans, V. C., Barker, G., Heesom, K. J., et al. (2012). De novo derivation of proteomes from transcriptomes for transcript and protein identification. *Nature Methods, 9* (12), 1207–1211.
12. Faulkner, S., Dun, M. D., & Hondermarck, H. (2015). Proteogenomics: Emergence and promise. *Cellular and Molecular Life Sciences, 72* (5), 953–957.

13. Forbes, S. A., Beare, D., Gunasekaran, P., et al. (2015). COSMIC: Exploring the world's knowledge of somatic mutations in human cancer. *Nucleic Acids Research, 43* (Database issue), D805–D811.
14. Hanahan, D., & Weinberg, R. A. (2000). The hallmarks of cancer. *Cell, 100* (1), 57–70.
15. Jumeau, F., Com, E., Lane, L., et al. (2015). Human spermatozoa as a model for detecting missing proteins in the context of the chromosome-centric Human Proteome Project. *Journal of Proteome Research, 14* (9), 3606–3620.
16. Kandoth, C., McLellan, M. D., Vandin, F., et al. (2013). Mutational landscape and significance across 12 major cancer types. *Nature, 502* (7471), 333–339.
17. Krasnov, G. S., Dmitriev, A. A., Kudryavtseva, A. V., et al. (2015). PPLine: An automated pipeline for SNP, SAP, and splice variant detection in the context of proteogenomics. *Journal of Proteome Research, 14* (9), 3729–3737.
18. Kumar, P., Henikoff, S., & Pauline, C. N. (2009). Predicting the effects of coding non-synonymous variants on protein function using the SIFT algorithm. *Nature Protocols, 4* (8), 1073–1082.
19. Kurian, A. W., Hare, E. E., Mills, M. A., et al. (2014). Clinical evaluation of a multiple-gene sequencing panel for hereditary cancer risk assessment. *Journal of Clinical Oncology, 32* (19), 2001–2009.
20. Lander, E. S., Linton, L. M., Birren, B., et al. (2001). Initial sequencing and analysis of the human genome. *Nature, 409* (6822), 860–921.
21. Landrum, M. J., Lee, J. M., Riley, G. R., et al. (2014). ClinVar: Public archive of relationships among sequence variation and human phenotype. *Nucleic Acids Research, 42* (Database issue), D980–D985.
22. Lappalainen, I., Lopez, J., Skipper, L., et al. (2013). DbVar and DGVa: Public archives for genomic structural variation. *Nucleic Acids Research, 41* (Database issue), D936–D941.
23. Legrain, P., Aebersold, R., Archakov, A., et al. (2011). The human proteome project: Current state and future direction. *Molecular and Cellular Proteomics, 10* (7), M111.009993.
24. McDermott, U., Downing, J. R., & Stratton, M. R. (2011). Genomics and the continuum of cancer care. *New England Journal of Medicine, 364* (4), 340–350.

25. McLaren, W., Pritchard, B., Rios, D., et al. (2010). Deriving the consequences of genomic variants with the Ensembl API and SNP effect predictor. *Bioinformatics, 26* (16), 2069–2070.
26. Meyerson, M., Gabriel, S., & Getz, G. (2010). Advances in understanding cancer genomes through secondgeneration sequencing. *Nature Reviews Genetics, 11* (10), 685–696.
27. Nagaraj, N., Wisniewski, J. R., Geiger, T., et al. (2011). Deep proteome and transcriptome mapping of a human cancer cell line. *Molecular Systems Biology, 7,* 548.
28. Nagaraj, S. H., Waddell, N., Madugundu, A. K., et al. (2015). PGTools: A software suite for proteogenomic data analysis and visualization. *Journal of Proteome Research, 14* (5), 2255–2266.
29. Nesvizhskii, A. I. (2014). Proteogenomics: Concepts, applications and computational strategies. *Nature Methods, 11* (11), 1114–1125.
30. Pabinger, S., Dander, A., Fischer, M., et al. (2014). A survey of tools for variant analysis of next-generation genome sequencing data. *Briefings in Bioinformatics, 15* (2), 256–278.
31. Paik, Y. K., & Hancock, W. S. (2012). Uniting ENCODE with genome-wide proteomics. *Nature Biotechnology,* 30 (11), 1065–1067.
32. Prieto, G., Aloria, K., Osinalde, N., et al. (2012). PAnalyzer: A software tool for protein inference in shotgun proteomics. *BMC Bioinformatics, 13*, 288.
33. Reiter, L., Claassen, M., Schrimpf, S. P., et al. (2009). Protein identification false discovery rates for very large proteomics data sets generated by tandem mass spectrometry. *Molecular and Cellular Proteomics, 8* (11), 2405–2417.
34. Segura, V., Medina-Aunon, J. A., Mora, M. I., et al. (2014). Surfing transcriptomic landscapes. A step beyond the annotation of chromosome 16 proteome. *Journal of Proteome Research, 13* (1), 158–172.
35. Smigielski, E. M., Sirotkin, K., Ward, M., et al. (2000). dbSNP: A database of single nucleotide polymorphisms. *Nucleic Acids Research, 28* (1), 352–355.
36. Sotiriou, C., & Pusztai, L. (2009). Gene-expression signatures in breast cancer. *New England Journal of Medicine,* 360 (8), 790–800.
37. Tabas-Madrid, D., Alves-Cruzeiro, J., Segura, V., et al. (2015). Proteogenomics dashboard for the Human Proteome Project. *Journal of Proteome Research,* 14 (9), 3738–3749.

38. Tamborero, D., Gonzalez-Perez, A., Perez-Llamas, C., et al. (2013). Comprehensive identification of mutational cancer driver genes across 12 tumor types. *Science Reports, 3*, 2650.

39. Tomczak, K., Czerwińska, P., & Wiznerowicz, M. (2015). The Cancer Genome Atlas (TCGA): An immeasurable source of knowledge. Contemporary Oncology (Pozn), 19 (1A), A68–A77.

40. Trapnell, C., Hendrickson, D. G., Sauvageau, M., et al. (2013). Differential analysis of gene regulation at transcript resolution with RNA-Seq. *Nature Biotechnology, 31* (1), 46–53.

41. Venter, J. C., Adams, M. D., Myers, E. W., et al. (2001). The sequence of the human genome. *Science, 291* (5507), 1304–1351.

42. Vizcaíno, J. A., Côté, R. G., Csordas, A., et al. (2013). The PRoteomics IDEntifications (PRIDE) database and associated tools: Status in 2013. *Nucleic Acids Research, 41* (Database issue), D1063–D1069.

43. Wang, X., & Zhang, B. (2013). customProDB: An R package to generate customized protein databases from RNA-Seq data for proteomics search. *Bioinformatics, 29* (24), 3235–3237.

44. Woo, S., Cha, S. W., Na, S., et al. (2014). Proteogenomic strategies for identification of aberrant cancer peptides using large-scale next-generation sequencing data. *Proteomics, 14* (23–24), 2719–2730.

45. Yang, X., & Lazar, I. M. (2014). XMAn: A Homo sapiens mutated-peptide database for the MS analysis of cancerous cell states. *Journal of Proteome Research, 13* (12), 5486–5495.

46. Zhang, B., Wang, J., Wang, X., et al. (2014). Proteogenomic characterization of human colon and rectal cancer. *Nature, 513* (7518), 382–387.

CHAPTER 7

PROTEOGENOMICS ANALYSIS OF ONCOGENIC MUTATIONS

CONTENTS

One of the feasible applications of proteogenomics is the analysis of oncogenic mutations. In this chapter, we demonstrate how the proteogenomics workflow is executed in one of the major cancer types worldwide. Firstly, we give an illustration of lung cancer types and mutations and then, we show how proteogenomics could be a highly beneficial tool.

7.1. INTRODUCTION

Worldwide, Lung cancer is the most prevalent sort of cancer and it is a foremost cause of death (Jemal et al., 2011; the data base of Japanese Ministry of Health and Labor and Welfare, 2013). In Japan, annual deaths from lung cancer are growing and presently reach around 70,000 (The data base of Japanese Ministry of Health and Labor and Welfare, 2013), while in the United States with a current declining trend in mortality, more than 160,000 surrender per year (The data base of National Cancer Institute at the National Institute of Health, 2013). Tobacco smoking is the chief risk factor for lung cancer; yet, around 25% of lung cancers worldwide take place in non-smokers (Ferlay et al., 2010; Siegel et al., 2011). Furthermore, the risk of lung cancer varies by ethnicity/race. In the United States, almost 10% of patients with lung cancer are smokers (Scagliotti et al., 2009), while in Asia, >30% of lung cancer patients are non-smokers and more than 50% of lung cancers occur in non-smoker women (Toh et al., 2004).

Weinstein (2001) suggested that cancer cells gain aberrations in numerous oncogenes and tumor suppressor genes but that the inactivation of a sole pivotal oncogene can encourage cancer cells to differentiate into cells with a normal phenotype or to suffer apoptosis (Oncogene Addiction). Only scarce motivator oncogenic genes would lead to their dysfunctions and result in anti-tumor properties while plentiful abnormal oncogenes are expressed in a tumor. The successful implementation of this phenomenon is the targeted therapy targeting lung cancers with mutated epidermal growth factor receptor (EGFR) by EGFR-tyrosine kinase inhibitors (EGFR-TKIs), erlotinib and gefitinib, which subsequently altered the treatment strategy of lung cancer.

Numerous clinical trials, such as WJTOG3405 (Mitsudomi et al., 2010) and the NEJ002 (Miyauchi et al., 2015) studies, described the advantage of gefitinib over carboplatin (CBDCA) and paclitaxel (PTX; Taxol) (CBDCA/PTX) on clinical end point as progression-free survival (PFS) as the first-line treatment for advanced non-small cell lung cancer (NSCLC) possessing an epidermal growth factor receptor (EGFR) mutation. Furthermore, in 2012 the molecular target drug, crizotinib, was permitted for lung cancers with an anaplastic lymphoma kinase (ALK) gene translocation (the fusion gene, EML4-ALK) (Camidge and Doebele, 2012) Furthermore, those targeting lung cancers addicted RET (Sasaki et al., 2012), ROS1 (Shaw et al., 2012), and HER2 (Mar and Vredenburgh, 2015) have also been undergoing further development. It appears that such personalized managements are altering lung cancer to a chronic disease, which might even be able to be healed.

7.2. LUNG CANCER SUBTYPES AND MUTATIONS

7.2.1. Lung Cancer Subtypes

One of the subtypes in which cancerous large cells grow deprived of any characteristic tissue construct is Large-cell lung cancer (LCC). Whereas Small-cell lung cancer (SCLC) is the subtype of an aggressive neuroendocrine tumor containing small bare nuclei cells. Travis et al. (1991) suggested a novel subtype of large-cell lung carcinoma, called large cell neuroendocrine carcinoma (LCNEC) in 1991, and the World Health Organization lastly approved it for the revised pathological sorting of lung cancer in the 1999 and 2004 WHO classifications. (Travis et al., 1999, 2004) LCNEC displays morphology analogous to LCC, but neuroendocrine differentiation like SCLC that could be mediated by the expression of at least one of three illustrative neuroendocrine proteins; synaptophysin (Syn), chromogranin A (CGA) or CD56. The developmental tracking of the tissue origin is presently unidentified for these three sorts of lung cancers.

LCNEC shows a poor prognosis, likewise to small-cell lung carcinoma (SCLC), and the survival rate is only 18% in IA-stage, just by resection (Dresler et al., 1997; Battafarano et al., 2005). Presently, equally to non-small-cell lung carcinoma (NSCLC), resection is the principal choice, tailed by adjuvant therapy, as for SCLC. Surgical resection of LCNEC in various series has been illustrated with 5-year actuarial survival that is worse than that described for other histological subtypes of non-small-cell lung cancer (NSCLC). There have been substantial arguments on whether these tumors should be categorized and cured as NSCLC or SCLC. The wide scale epidemiologic study has related the presenting and prognostic features of patients with LCNEC with those of patients with SCLC or other large cell carcinomas (OLCs) with regard to lung cancer-specific survival (LCSS) or overall survival (OS) rates for patients experiencing decisive resection without radiotherapy (S-NoRT), and they have settled that LCNEC should endure being categorized and managed as a large cell carcinoma because the histopathologic, clinical and biologic features of LCNEC are more analogous to OLC than to SCLC (Varlotto et al., 2011; Travis, 2010).

7.2.2. Lung Adenocarcinoma Classification

In a growing tendency worldwide, progresses in chest high-resolution computed tomography (HRCT) scanning technology have allowed for the localization of small adenocarcinoma nodules (Nakamura and Saji, 2014) at a prior and possibly more treatable stage of development than previously probable (Koike et al., 2009). There are 90 million present and ex-smokers in the United States who are at augmented risk of lung cancer. The published data from the National Lung Screening Trial (NLST) propose that annual screening with low dose thoracic CT scans in heavily smoking people can decrease lung cancer mortality by 20% and all-cause mortality by 7% (The National Lung Screening Trial Research Team, 2011).

In 2011, the novel pathologic cataloging of lung adenocarcinoma was suggested by the International Association for the Study of Lung Cancer (IASLC), the European Respiratory Society (ERS) and the American Thoracic Society (ATS) (Travis et al., 2011). In the new assortment, the idea of adenocarcinoma in situ (AIS) and minimally invasive adenocarcinoma (MIA) were recently presented and the concept of bronchioloalveolar carcinoma (BAC) was eliminated. In addition, invasive adenocarcinomas were categorized into 6 subtypes, papillary (PAP), micropapillary (MP), lepidic (LEP), acinar (CAN), solid (SLD), and variants, depending on the predominant histologic pattern. Both AIS and MIA were distinct as tumors $\leq$ 3 cm in size. LEP is an invasive adenocarcinoma showing former nonmucinous BAC pattern with $>$ 5 mm invasion. AIS is a preinvasive lesion possessing pure lepidic growth without invasion. MIA is also lepidic predominant tumor but with $\leq$ 5 mm invasion.

These 3 lepidic sort adenocarcinomas are guessed to possess escalating progression from AIS, MIA, to LEP. After the whole resection of AIS or MIA, frequently 100% of recurrence free 5year survival can be attained (Travis et al., 2011), while some recurrent cases existed after resection of LEP (Yoshizawa et al., 2011, 2013; Gu et al., 2013). As postoperative prognoses between the AIS plus MIA group and LEP are dissimilar, differential protein expressions linked to invasiveness of cancer cells in each subtype should execute significant roles to define local recurrences and survivals. Lately the motivating mutations, including the BRAF, ALK,

EGFR, KRAS, HER2, RET, and PIK3CA, were explored across the IASLC/ATS/ERS morphologic categorizations in lung adenocarcinoma integrated with the clinicopathological features to appraise their mutual correlation, in which lung adenocarcinomas acquired from 1015 Chinese patients were analyzed (Hu et al., 2014). The motivating mutations, BRAF (exons 11–15), EGFR (exons 18–22), HER2 (exons 18–21), KRAS (exons 2–3) and PIK3CA (exons 9–20) were amplified by employing the polymerase chain reaction (PCR) with cDNA used for Sanger sequencing.

The RET and ALK rearrangements were screened by utilizing PCR and quantitative real-time PCR with cDNA and established with fluorescence in situ hybridization in formalin-fixed paraffin embedded (FFPE) tissue specimens. Figure 7.1 illustrated that all driver mutations among the IASLC/ATS/ERS organizations were equally exclusive except in 18 patients possessing simultaneous EGFR and PIK3CA mutations, and 1 with both the RET and PIK3CA mutations and 4 with both the KRAS and PIK3CA mutations. The EGFR mutation was 64.7% in overall frequency, extremely upper than in the Caucasian population, and the KRAS mutation was 7.1%, much inferior to in Caucasian patients.

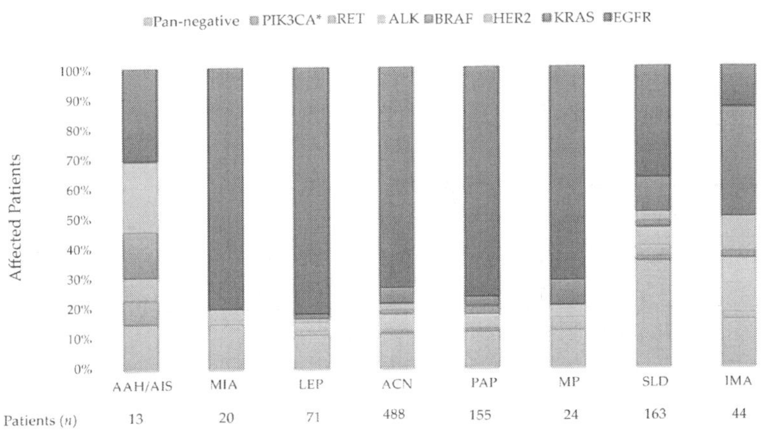

Figure 7.1. Hinge mutation spectrum, according to the novel IASLC/ATS/ERS classification (Hu et al., 2014).

Non-smoker East Asian females have an inclination to develop adenocarcinoma, and these non-smokers show higher treatment reaction rates to epidermal growth factor receptor tyrosine kinase inhibitors (EGFR-TKIs), such as erlotinib (Tarceva™) and gefitinib (Iressa™), than those with a history of tobacco smoking (Ha et al., 2015). EGFR-TKIs inhibit EGFR phosphorylation and succeeding signal transduction pathways accused for proliferation, metastasis, angiogenesis and apoptosis inhibition, and gefitinib is the leading molecular targeting drug considerably beneficial for Asian lung cancer patients. Figure 7.2 revealed the frequency of driver gene mutations in lung adenocarcinomas from East Asian non-smoker females, among which EGFR mutations were the most commonly existing mutation in lung adenocarcinomas of female non-smokers.

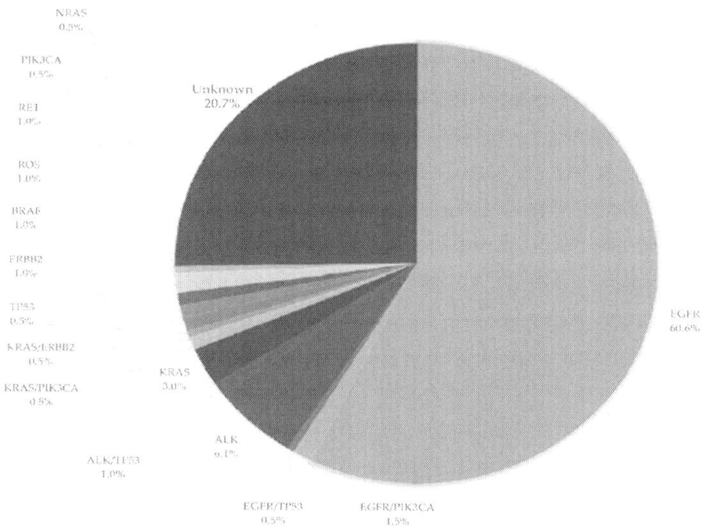

Figure 7.2. Frequency of hinge gene mutations in lung adenocarcinomas from East Asian never-smoker females (Ha et al., 2015).

Mutations in the TK domain of the EGFR were recognized in those patients with intractable non-small cell lung cancer who attained dramatic tumor reactions to gefitinib. Though it is not well assumed yet why NSCLCs takes place in non-smokers, it has now been recognized that non-smoker-related NSCLCs involve diseases discrete biologically to smoking-related NSCLCs: The first are characterized by the substantially high occurrence in EGFR mutations, the last show more common KRAS mutations (Oxnard et al., 2014; Sun et al., 2007). The major prevalence of motivating oncogenes in lung adenocarcinomas from non-smokers definitely leads diseases putative oncogene- driven malignancies in these diseases, in which the management by kinase inhibitors helps patients effectively.

7.3. CLINICAL PROTEOGENOMICS

Lots of clinical samples are attained every day for procedures in diagnostic tests that backs clinical decision making. These clinical samples (tissues, biopsies, blood, *etc.*) can also be recorded into repositories for usage in future studies examining the etiology of diseases using omics methods. Consequently, infrastructure accumulation of standardized biobanking is progressively necessary within the clinical omics community since the samples themselves have intrinsic standards in the judgement of outcomes of clinical trials (Végvári et al., 2011a, b; Marko-Varga, 2011; Marko-Varga et al., 2011; Malm et al., 2012; LaBaer, 2012). The samples can be recovered from pathology laboratories with the endorsement from ethical committees of medical institutes and hospitals. Numerous sorts of disease specimens are present, such as frozen and FFPE tissues; biopsies; and body fluids comprising blood, serum, plasma, and urine; interstitial fluid; cyst material; ascites fluid; and pancreatic juice.

7.3.1. Laser Microdisection and Protein Solubilization

In hospitals and reseach institutes, tumor tissues acquired by surgical resection are classically fixed in 4% paraformaldehyde and regularly processed for paraffin sectioning. Cancerous lesions can be recognized on serial tissue sections stained with hematoxylin and eosin (HE). Figure 7.3 demonstrates (A) focal ground-glass opacity on chest HRCT, which lesions are recognized as AIS, MIA, or LEP and (B) a representative HE-stained image of LEP. Laser microdissection (LMD) makes it probable to assemble target cells from a variety of FFPE cancer tissues.

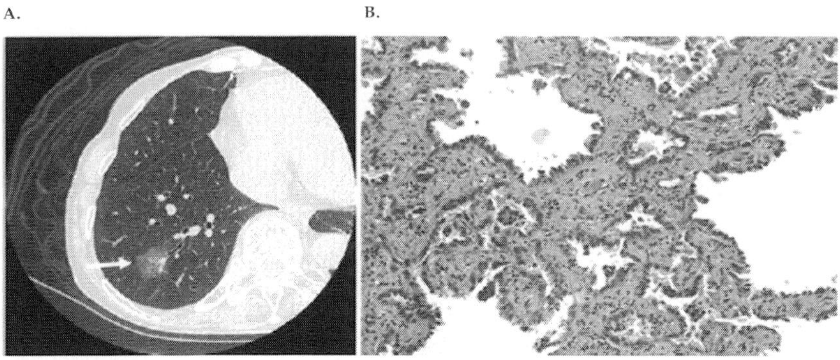

Figure 7.3. (a) Focal ground-glass opacity, which lesions are identified as AIS, MIA, or LEP. (b) A representative hematoxylin and eosin-stained image of LEP

For shotgun proteomic analysis, 10-μm sections set from the same tissue block are attached onto DIRECTOR™ slides (OncoPlexDx, Rockville, MD, USA), deparaffinized double with xylene for 5 min, then rehydrated with graded ethanol solutions and distilled water, and later stained by hematoxylin (Prieto et al., 2005; Hood et al., 2005; Kawamura et al., 2010; Nomura et al., 2011). Slides are air dried and exposed to LMD with a Leica LMD7000 (Leica Micro-systems GmbH, Ernst-Leitz-Strasse, Wetzlar, Germany). Classically, ca. 30,000 cells (ca. 8 mm 2) per tissue sample are transported directly to a 1.5-mL low-binding plastic tube. Figure 7.4 demonstrates the hematoxylin-stained LEP tissue before and after LMD (C-1 and C-2, respectively).

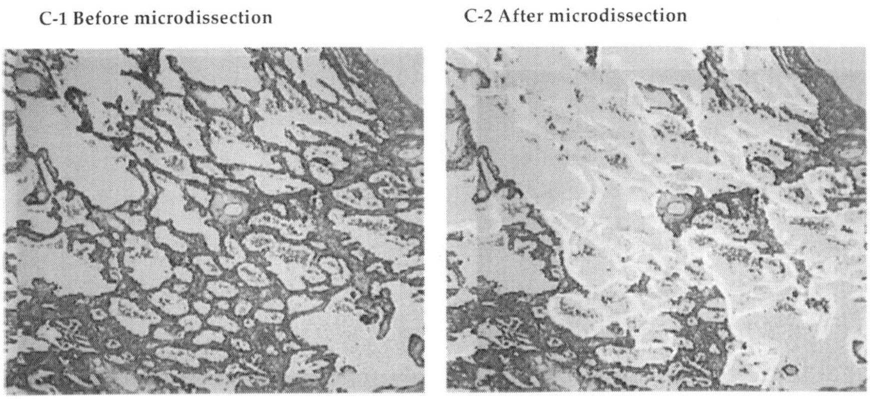

Figure 7.4. The hematoxylin-stained LEP tissue before and after laser microdissection (C-1 and C-2, respectively).

Proteins/peptides from dissected cells can be removed by employing numerous protocols (Prieto et al., 2005; Wisniewski et al., 2011). For example, according to the protocol of a Liquid Tissue™ MS Protein Prep kit (Expression Pathology) (Prieto et al., 2005), the cellular material, suspended in the liquid tissue buffer, is incubated at 95°C for 90 min, cooled on ice (3 min), and then enzymatically digested, tailed by reduction and alkylation. The liquid tissue digests can be kept at −20°C till proteomic analysis.

7.3.2. Protein Identification and Quantification

Current progress in MS could make proteomics open to detailed investigative and targeted quantitative analysis of proteins expressed in a composite clinical specimen (Marko-Varga et al., 2011; Nakamura et al., 2012). MS is significantly profitable due to its particularly high competence of capturing/ recognizing/sequencing of proteins/ peptides expressed in a complex clinical specimen, with high sensitivity and high accuracy, dissimilar to others (Nishimura and Tojo, 2014).

An investigative proteomic analysis usually encompasses extraction and/or direct tryptic digestion of entire expressed proteins in a composite biological sample, and then the peptide mixture attained is exposed to liquid chromatography (LC)/electrospray ionization-tandem MS "ShotGun" analytical platform so as to sequence these by searching against protein sequence databases. Protein recognition in shotgun proteomic approaches (bottom- up) can be now executed by four peptide sequencingstrategies using MS/MS spectra: (A) database search, (B) spectral library matching, (C) hybrid approaches using sequence-tag determination followed by database search, and (D) de novo sequencing as illustrated in Figure 7.5. Several hundred to several thousand (more than 10,000 in some cases) different protein species can typically be identified in such exploratory clinical proteomic studies (Michalski et al., 2011; Panchaud et al., 2011; Geiger et al., 2010; Gillet et al., 2012; Bern et al., 2010; Physikron Mass Spectrometry Systems, 2013; Gorshkov et al., 2015), in which label-free semi-quantitative comparison with statistical evaluation is mainly performed to elucidate proteins specifically relevant to a disease subtype.

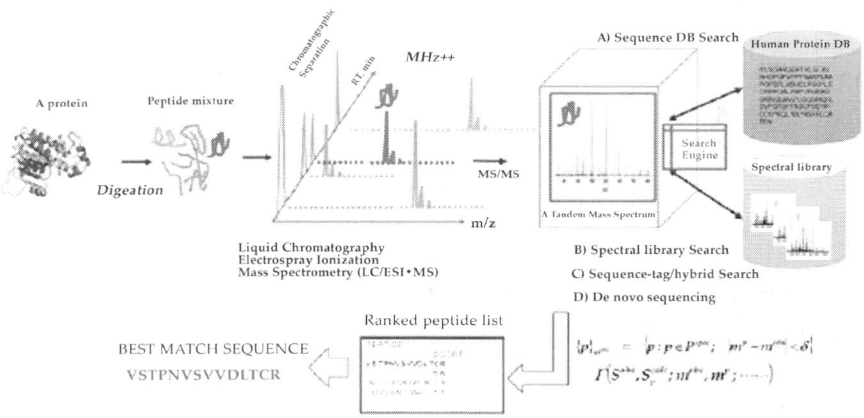

Figure 7.5. An illustration of exploratory mass spectrometry (MS) -based proteomic analysis.

7.4. DEVELOPMENT OF PROTEOGENOMICS TECHNOLOGIES TO IDENTIFY/QUANTIFY MUTATED PROTEINS

7.4.1. Exploratory In-depth Analysis

Proteins are functionally dynamic since these biological molecules display numerous forms of sequences, comprising not only post-translational modifications/truncations/variants but also mutations/ rearrangements. For recognition of such dynamic proteins from clinical samples, the advance of proteome bioinformatics is of growing significance. There are presently numerous gene–protein databases, which, nevertheless, endure to be merged. The total count of protein coding genes is stated to be ~20,000, from which the major proteins are estimated to be around ca. 38,400 in neXtProt and 68,000 in UniProt. These databases have been most frequently utilized for protein recognition in classical proteomic studies. Nevertheless, there are other databases not publicly accessible such as the fusion-gene database (ca. 10,000 species), the MuPdb mutation database (ca. 1,200,000 species) and the "Short" protein (the number of amino acids <100) database (ca. 6500 species), which have not been entirely annotated so far. MuPdb comprises all potential non-synonymous single nucleotide polymorphism (nsSNP) sequences identified from genomic studies (Ensembl database).

MS-based proteomics produces high-quality sequential protein data on the human proteome. Nevertheless, the shortage of a comprehensive database comprising a whole assortment of nsSNP products and fusion genes inhibits recognition of mutant and fusion proteoforms even if high-quality mass spectra are accessible. A current study to define the distribution of known oncogenic motivating mutations in female non-smoker Asian patients with lung AC demonstrated that around 79% of patients possessed driver gene mutations as described before (Ha et al., 2015). While multiplex gene analysis does not look to be representative, MS-based multiplex assays can be executed by querying proteome MS/MS datasets attained from clinical samples against the diversity of databases, which will very likely recognize novel functional proteins significant to therapeutic targets that have continued to be unidentified until now (Figure 7.5).

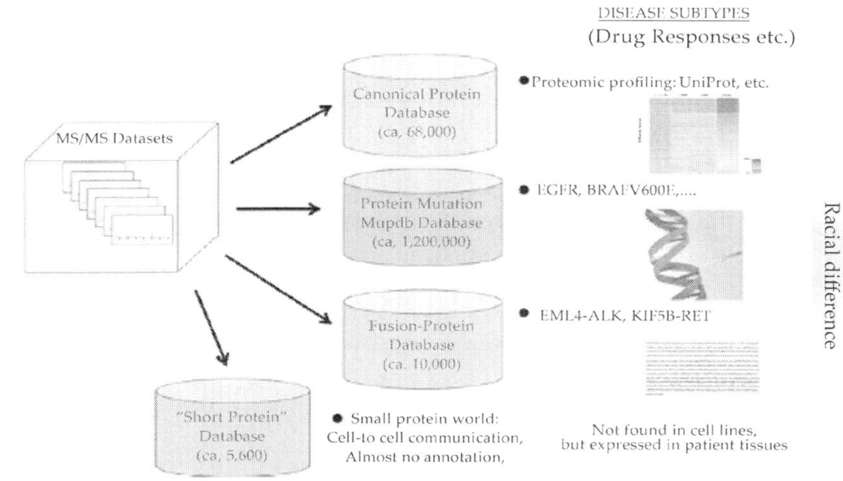

Figure 7.6. Proteogenomics analysis of clinical proteome datasets by searching for various protein/gene databases.

7.4.2. Quantitative Identification of Oncogenic Fusion Gene Products

Gene fusion occasions characterize the most frequent type of genomic rearrangement as a result of inversions, interstitial deletion, or translocations. The leading fusion gene, BRC-ABL1 was revealed by Peter Nowell & David Hungerford (1960) and has been a diagnostic genetic marker for chronic myelogenous leukemia in addition to be the drug target of imatinib (Gleevec™). The detection of TMPRSS2-ERG fusions in prostate cancer (Tomlins et al., 2008) and EML4-ALK fusion in NSCLC tumors (Soda et al., 2007) proposed a reasonably common occurrence of gene fusion events in solid tumors, which produce new oncogenic fusion proteins. These fusion gene products are chiefly limited to tumor cells, which would be convenient diagnostic and therapeutic targets.

The fusion protein FusPdb database, constructed from the integrated chimeric gene database, ChiTaRS (2013), presently comprises around 9300 candidates. Figure 7.7 demonstrates the workflow of recognizing and quantifying fusion gene products, in which the NPM-ALK fusion protein (680 aa) generated by the gene translocation t(2;5) (p23;q35) is displayed as an example. Current developments in mass spectrometric make it likely to recognize various fusion junction peptides/fusion proteins by LC–MS/MS in high-resolution and high sensitivity approaches, in which subsequent datasets are questioned against the FusPdb/canonical protein databases and/ or directly compared to the ChiTaRS database using suitable searching software, such as COMET (Eng et al., 2015).

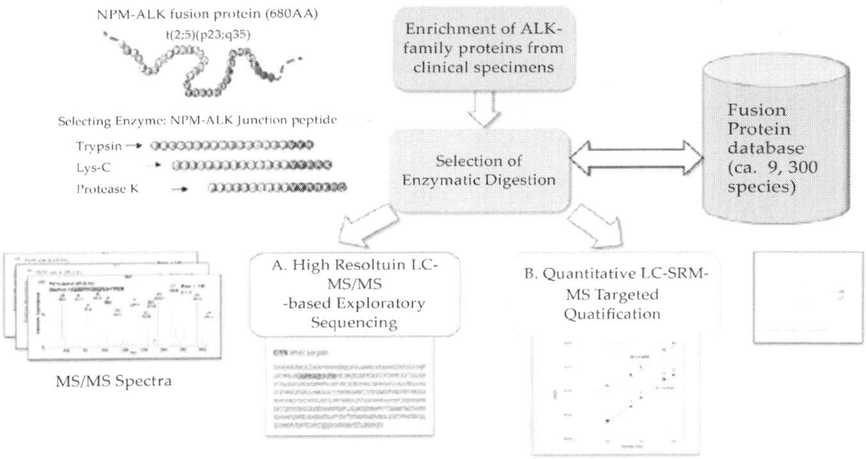

Figure 7.7. A workflow of identifying and quantifying fusion gene products, in which the NPM-ALK fusion protein (680 aa) produced by the gene translocation t(2;5)(p23;q35) is shown as an example.

Preceding to the MS-based analysis of fusion gene products of concern, it is significant to develop approaches for affinity-based enrichment from a composite clinical sample, tailed by efficient digestion. Junction peptides can be planned using the FuseProt database, which is all probable peptides containing the junction position of a fusion protein. For MS-based quantification of targeted fusion-proteins of concern, exclusive junction peptides comprising junction points are planned by choosing suitable enzymatic proteases, in which their enzymatic miss-cleavages are also put into consideration. MS-based targeted assays would classically be executed using the acquisition methods of the single-ion monitoring (SIM) and selected-reaction monitoring (SRM), where absolute quantification can be attained with spiked authentic stable isotope-labeled junction peptide (AQUA junction peptide) of an identified amount. There are the main three variations of EML4-ALK fusion gene products. Figure 7.8 shows (A) the

11 SRM-transition peaks acquired for the authentic EML4-ALK variant-1 junction peptide, k.YIMSNSGDYEIYL-YRRK.h, being generated via Lys-C digestion and (B) the SIM quantitation of the authentic EML4-ALK Variant-3 junction peptide, k.NSQ-VYRRK.h (a Lys-C digestion product) of around 10–1000 fmol spiked into a plasma matrix (around 100 ng/µl). Both were measured on a Q-Exactive high-resolution mass spectrometer. The MS-based targeted quantification assay of numerous fusion gene products comprising their variants is precise and stretchy, by which numerous targeted fusion gene products can be quantitatively checked instantaneously with tissue/biopsy of a patient.

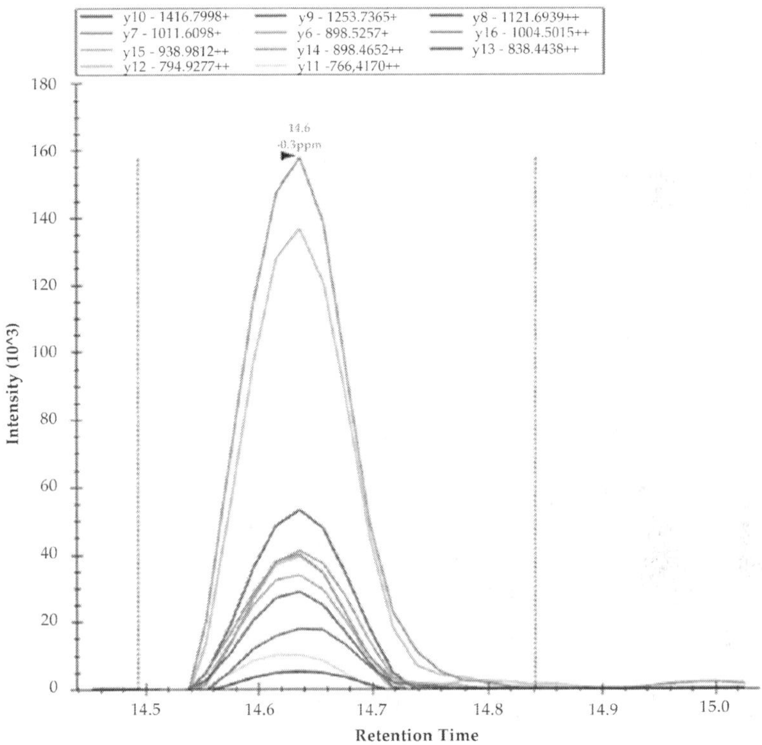

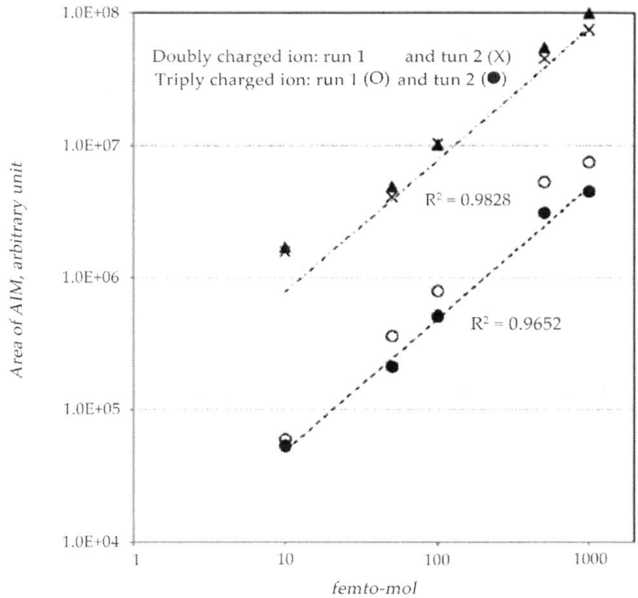

Figure 7.8. The 11 SRM-transition peaks obtained for the authentic EML4-ALK variant-1 junction peptide.

For instance, instantaneous MS-based quantitation of both EML4-ALK and KIF5B-RET—a multiplex assay—can be planned as follows: (1) targeted fusion proteins (comprising variants) can be improved using suitable antibodies, (2) relevant enriched samples are digested by specific proteases, and (3) the corresponding peptide mixture samples are joined into one sample, which is exposed to targeted MS-based quantification. An MS-based multiplex fusion protein assay has a high prospective to enable the formation of the conclusive diagnosis, helping the patient by allowing administration of prime treatment, and comes with a little assay cost, though additional expansion of the technology is essential.

7.5. CELLULAR PATHWAYS AFFECTED BY SOMATIC MUTATIONS

The identification of recurring mutations in EGFR and fusions concerning ALK and other receptor tyrosine kinases has significantly changed the standard of treatment of patients with lung AC. Existing guidelines endorsed the molecular genotyping of AC to regularly comprise the EGFR and ALK status, changes which are found to happen in around 25% of patients with

AC who profit more from official targeted inhibitor therapies than from classical chemotherapy. Such somatic changes, mutations, and fusions in lung cancers commonly affect cellular pathway events involved in lung cancer subtypes. Figure 7.9 recapitulates cellular pathways, the events of which are affected by somatic changes in lung cancer subtypes, namely AC, squamous cell carcinoma, and SCLC (Shtivelman et al., 2014).

a

b

Pathway	AC	SQCC	SCLC
RTK	EGFR ALK MET, ERBB2, ROS, RET	EGFR, FGFR1-3 ERBB2, ERBB3 DDR2	FGFR1
RAS/RAF	KRAS, NF1, BRAF, NRAS	NF1, KRAS, HRAS NRAS RASA1, BRAF	
PI3K/AKT	PIK3CA, PTEN, AKT1	PIK3CA PTEN AKT1, AKT2 AKT3, TSC1-2	PTEN
LKB1/AMPK			
TP53	P53, MDM2	TP53	TP53
RB1/CDKNA2	CDKNA2	CDKNA2 RB1	RB1, CCNE1
MYC	MYC		MYC, MYCN, MYCl
Epigenetic reguration	SMARCA4 ARID1A, SETD2	MLL2	EP300, CREBBP, MLL
Developmental pathway	NKX2.1/TTF1	SOX2, TP83, NOTCH1, NOTCH2, ASCL4, FOXP1	SLIT2, EPHA7
Oxidative stress response	KEAP1	KEAP1, NRF2, CUL3	

Figure 7.9. (a) Cellular pathways affected by somatic alterations and (b) genes involved in each type of lung cancer are listed in order of the frequency of alterations found.

The most vital information is how proteins expressed significantly in disease subtype interchange with other vital proteins and pathways to assess biomarker candidates and therapeutic targets. Numerous open PPI databases are obtainable; present versions comprise Reactome (Reactome Pathway database, 2016) and BioGRID (The Biological General Repository for Interaction Datasets, 2016), and PPI network analysis can be executed by selected network construction algorithms, using, for example, the Search Tool for the Retrieval of Interacting Genes/Proteins (STRING) database (STRING 10 2016) and the Cytoscape (Cytoscape3 2014), a software environment for incorporated models of biomolecular interaction networks. PPI networks clarified so far contain nodes and edges, where nodes are proteins experimentally recognized, and edges are the expected functional associations based on primary databases including KEGG and GO (2016), the principal literature, and so on. Therefore, it has become likely to clarify protein networks relevant to a disease subtype employing its proteomic datasets. The notion of biomarkers has been changing from classical biomarkers —single proteins—to certain protein networks or dynamically changing networks (Wu et al., 2014), since diseases can be looked at also as kinetic network disorders. PPI networks dynamically activated or deactivated in a disease subtype of concern would be directly linked to its molecular mechanisms in charge, which would result in the discovery of therapeutic. Figure 7.10 demonstrates the STRING PPI networks constructed from the LEP proteome datasets acquired from a clinical proteomic analytical study of cancerous cells laser-microdissected from lung cancer FFPE-tissue specimens (Kato et al., 2015).

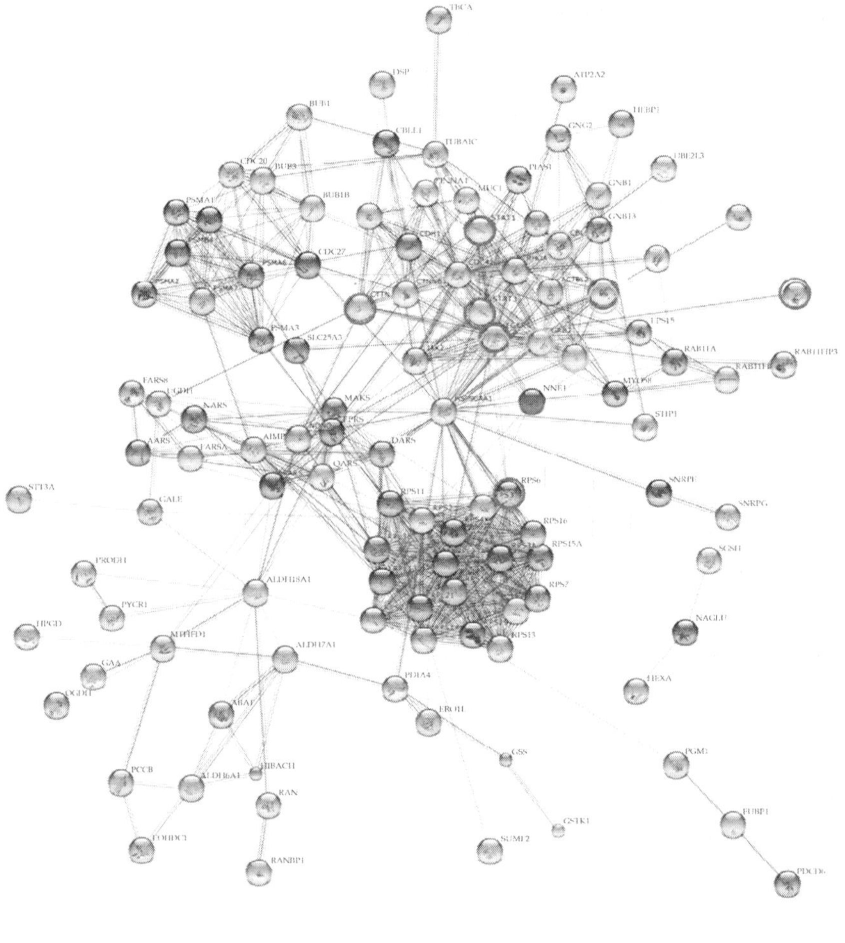

Figure 7.10. The STRING protein–protein interaction networks developed from the LEP proteome datasets.

The ErbB signaling pathway is linked to numerous cancer pathways. The ErbB family represents epidermal growth factor receptors, which execute a critical role in tumor growth. Overexpression of EGFR takes place in about 60% NSCLCs, with patients with AC harboring the highest frequency (Shtivelman et al., 2014). Hypoxia-inducible factors (HIFs) control the transcription of genes that facilitate the reaction to hypoxia (reduced O 2 availability) (Semenza, 2010). Varied products of HIF-1 action such as induction of the Met protein, hepatocyte growth factor, tailed by Met receptor activation, may lead to the poor prognosis related to hypoxic tumors, which are certainly more aggressive than their well-oxygenated counterparts. Molecules contributing in the ErbB and HIF-1 signaling pathways are signified by orange and red circles in Figure 7.10, respectively (Kato et al., 2015). Figure 7.11 demonstrates the results of the STRING gene set enrichments for LEP, MIA, and AIS gained for the 24 cancer-related KEGG pathways, which were clarified with their significance rank $p < 0.05$ after correction by false discovery rate (FDR) (Kato et al., 2015). It was shown how functional contribution of expressed proteins changes dramatically during disease stages, mirroring the mechanisms of disease progression.

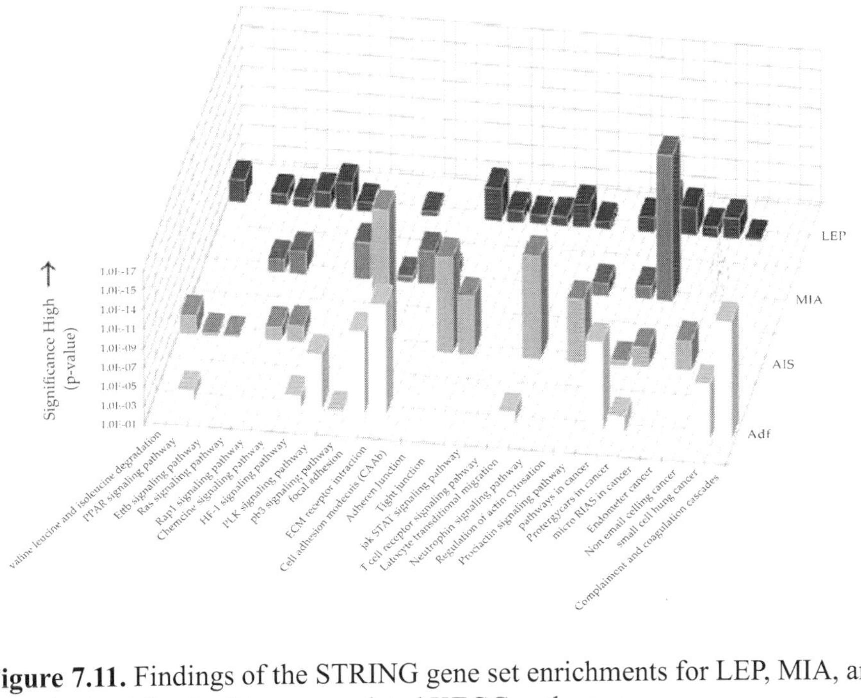

Figure 7.11. Findings of the STRING gene set enrichments for LEP, MIA, and AIS obtained for the 24 cancer-related KEGG pathways.

Therefore, MS-based exploratory proteomics using clinical specimens is a potential analytical platform, which makes it probable to disclose molecular networks significant to a disease subgroup, drug responders or nonresponders, good or poor prognosis, drug resistance, and so on. It should be highlighted that both somatic mutations and cellular pathways in disease subtypes are reciprocally strongly associated, and so both are required to be disclosed to understand molecular mechanisms of a disease subtype.

7.6. DRUG RESISTANCE AND GATEKEEPER MUTATIONS

7.6.1. Third-Generation EGFR Tyrosine Kinase Inhibitor (TKI) Drugs

First-generation EGFR-TKIs include Iressa™ (gefitinib) and Tarceva™ (erlotinib), which target activated EGFR mutations, comprising the L858R point mutation in Exon 21 and small deletions around E746-A750 in Exon 19, in the tyrosine kinase domain. Such mutations are frequently seen in lung adenocarcinomas (ACs) in East Asian populations. These two drugs display notable therapeutic properties in 40–50% of patients with lung cancer with AC in China, Korea, and Japan (Mitsudomi et al., 2010).

However, drug resistance has been reported to be linked to the administration of first generation EGFR-TKIs, where an attainment of the new T790M EGFR mutation in patients with EGFR mutation-positive lung cancer is viewed as the most common cause of resistance. Later, the second-generation EGFR-TKI Gilotrif™ (afatinib) (Hirsh, 2015; Bennouna and Moreno Vera, 2015) was presented; nevertheless, it seems that this TKI is no longer capable of binding due to the three-dimensional structural change of EGFR produced by the T790M mutation. Lately, the third-generation EGFRTKIs, AZD9291 (Jänne et al., 2015) and CO-1686 (rociletinib) (Sequist et al., 2015), established a high antitumor activity on non small-cell lung cancer (NSCLC), which had become resistant to first-generation EGFRTKIs; these TKIs have been established to selectively and irreversibly inhibit both EGFR with TKI- activated mutations and T790M and are presently in the expansion phase.

7.6.2. EGFR-TKI Drug Resistance and Dynamic Variation of Mutations

Lately, it was described that AZD9291 exhibited high antitumor activity in EGFR-T790M-positive patients with advanced NSCLC whose disease progressed with previous EGFR-TKI treatment and that rociletinib, which showed activity in both T790M (+) and (−) patients in a preclinical model, aids EGFR-mutation-positive patients with EGFR-T790M. The third-generation EGFR- TKI AZD9291 exhibited a remarkable result in that the median PFS was 13.1 months in a phase 2 trial restricted to T790M-positive cases (AURA trial, NCT01802632: AURA 2015). No difference was shown in the response rate between patients with Del19/T790M and L858R/T790M, which proposed that AZD9291 is specific to T790M (AURA 2015). The present diagnosis using gene analysis only judges whether the T790M mutation is present or not, and it is highly probable that a clone of T790M exists from the beginning.

Concerning tumor heterogeneity, three dissimilar scenarios of tumors harboring two EGFR mutations have been anticipated, in which activating and resistant mutations exist: (1) in cis on the same allele, (2) in trans on different alleles, or (3) in different clones (Figure 7.12) (Leone, 2013). Accordingly, it is important to develop an accurate methodology for T790M quantification, by which a correlation between the extent of T790M expression and drug efficacy can be examined, thus making it probable to describe a cutoff value for T790M abundance.

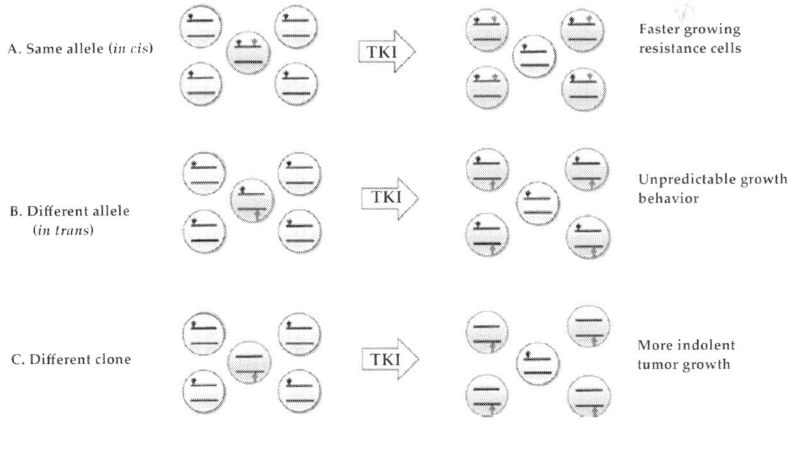

Figure 7.12 Three different cases of tumors harboring two EGFR mutations have been suggested, including activating and resistant mutations.

Figure 7.12 illustrates EGFR mutations in NSCLC, the crystal structure of the kinase domain of EGFR in complex with gefitinib (based on Protein Data Bank accession code 2ITY), and the location of the EGFR mutations (RCSB 2015).

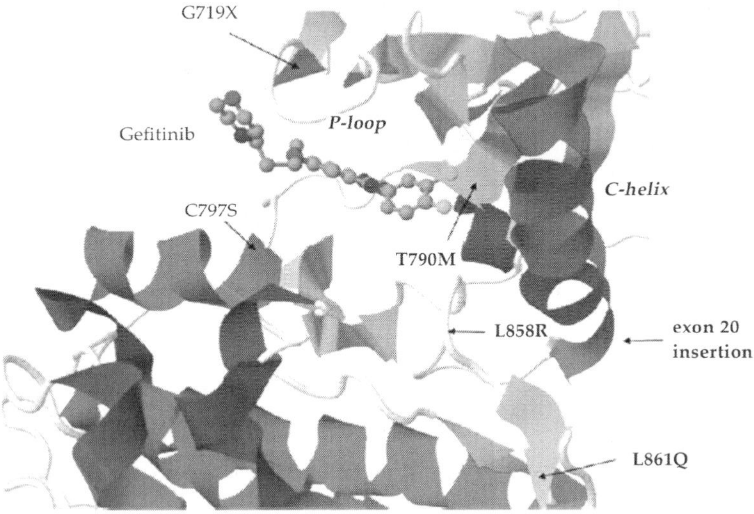

Figure 7.13 EGFR mutations in non small-cell lung cancer, the crystal structure of the kinase domain of EGFR in complex with gefitinib.

The efficacy of AZD9291 would also be restricted because resistance will be developed to this drug. C797 codon mutations in the EGFR tyrosine kinase-binding sites have been described as a resistance mechanism for AZD9291 in T790M. A current study proposed that there are both T790M (+) and (−) clones at baseline and that AZD9291 intervenes with the resistance of T790M (−) cells, although AZD9291 effectively subdues the growth of T790M (+) cells, which might be evaded by the activation of the HER2 and/or MET pathway. In addition, EGFR-TKI drug resistance has been linked to a new C797S mutation found near T790M (Thress et al., 2015) There are two reasonable patterns involved in AZD9291 resistance: (1) both T790M (+) and C797S (+) or (2) T790M (−) with circumventing taking place in a different signaling pathway such as HER2 or MET. Both resistance patterns might coincide within one type of tissue and within both primary and metastatic tumors. It has been described that gefitinib might be effective for C797S (+) cases (Ercan et al., 2015).

7.7. PERSPECTIVE

Both recognition and quantification of EGFR gatekeeper mutations, comprising mutation heterogeneity within a tissue, need to be executed using lung cancer tissue specimens received from patients to enhance the treatment for patients with EGFR mutation-positive NSCLC, including 1. recognition and quantitation data of targeted EGFR mutated proteins 2. Exploratory mass spectrometry (MS)-based clinical proteogenomics investigation of mutated proteins, comprising investigation of mutation heterogeneity within a tissue 3. Analysis of dynamic protein–protein interaction (PPI) networks of proteins considerably linked to a subgroup of patients with lung cancer with acquired resistance The most exciting investigation includes the pairwise MS-based proteogenomics analysis of EGFR mutation-positive (frozen and/or FFPE) tissues of lung cancer gained preceding to EGFRTKI treatment and after attainment of the drug resistance, which will expose comprehensive and direct molecular information on EGFR-TKI resistance.

In Japan, high-quality clinical specimens with comprehensive clinical and pathological data have been archived for numerous years within medical institutes and hospitals and comprise even early- stage cancers. In this period, it was demonstrated that drug efficacy varies by race (*e.g.*, Caucasians vs. Asians) due to oncogenic driver mutations/fusions particular to each race, represented by gefitinib and erlotinib. In comparison to time-consuming genomic analysis, MS-based proteogenomics approaches allow direct analysis of mutated and fusion proteins expressed in a clinical sample, which will deliver a powerful solution for the stratification of patients and drug discovery. When a discrete clinical study design utilizing valuable clinical samples, so as to say national possessions, is established and executed by teaming up meticulously with clinicians, an advanced treatment and drug discovery pipeline can be supplied from Japan, opening an access to Asia and its population of 3.9 billion.

REFERENCES

1. AURA. (2015). National Institutes of Health. AZD9291 First time in patients ascending dose study (AURA). 2015: Available from https://clinicaltrials.gov/ct2/ show/NCT01802632 . 16 Jan 2016.
2. Battafarano, R. J., Fernandez, F. G., Ritter, J., Meyers, B. F., Guthrie, T. J., Cooper, J. D., & Patterson, G. A. (2005). Large cell neuroendocrine carcinoma: an aggressive form of non-small cell lung cancer. *Journal of Thoracic and Cardiovascular Surgery, 130,* 166–172.
3. Bennouna, J., Moreno Vera, S. R. (2015). A fatinib-based combination regimen for the treatment of solid tumors: rationale, emerging strategies and recent progress. *Future Oncology*, 2015 Nov 25. doi: 10.2217/ fon.15.310. Available from http://www.futuremedicine.com/doi/abs/10.2217/fon.15.310 . 16 Jan 2016.
4. Bern, M., Finney, G., Hoopmann, M. R., Merrihew, G., Toth, M. J., & MacCoss, M. J. (2010). Deconvolution of mixture spectra from ion-trap data-independentacquisition tandem mass spectrometry. *Analytical Chemistry, 82*, 833–841.
5. BioGRID-The Biological General Repository for Interaction Datasets. Available from http://thebiogrid.org/. 15 Jan 2016.
6. Camidge, D. R., & Doebele, R. C. (2012). Treating ALK positive lung cancer-early successes and future challenges. *Nature Reviews Clinical Oncology, 9* (5), 268–277.
7. ChiTaRS (the database of chimeric transcripts and RNA Sequencing data database). (2013). Available from http://chitars.bioinfo.cnio.es/. 16 Jan 2016.
8. Cytoscape – An open source software platform for visualizing complex networks and integrating these with any type of attribute data. Available from http://www.cytoscape.org/. 15 Jan 2016.
9. Dresler, C. M., Ritter, J. H., Patterson, G. A., Ross, E., Bailey, M. S., & Wick, M. R. (1997). Clinical pathologic analysis of 40 patients with large cell neuroendocrine carcinoma of the lung. *The Annals of Thoracic Surgery, 63*, 180–185.
10. Eng, J. K., Hoopmann, M. R., Jahan, T. A., Egertson, J. D., Noble, W. S., & MacCoss, M. J. (2015). A deeper look into comet—Implementation and features. Journal of the American *Society for Mass Spectrometry, 26*, 1865–1874.

11. Ercan, D., Choi, H. G., Yun, C. H., Capelletti, M., Xie, T., Eck, M. J., Gray, N. S., & Jänne, P. A. (2015). EGFR mutations and resistance to irreversible pyrimidine-Based EGFR Inhibitors. *Clinical Cancer Research, 21* (17), 3913–3923.

12. Ferlay, J., Shin, H. R., Bray, F., Forman, D., Mathers, C., & Parkin, D. M. (2010). Estimates of worldwide burden of cancer in 2008: GLOBOCAN 2008. *International Journal of Cancer*, 127, 2893–2917.

13. Geiger, T., Cox, J., & Mann, M. (2010). Proteomics on an Orbitrap benchtop mass spectrometer using all-ion fragmentation. *Molecular & Cellular Proteomics, 9*, 2252–2261.

14. Gillet, L. C., Navarro, P., Tate, S., Röst, H., Selevsek, N., Reiter, L., Bonner, R., & Aebersold, R. (2012). Targeted data extraction of the MS/MS spectra generated by data-independent acquisition: A new concept for consistent and accurate proteome analysis. *Molecular & Cellular Proteomics, 11,* 1–17.

15. Gorshkov, V., Verano-Braga, T., & Kjeldsen, F. (2015). SuperQuant: A data processing approach to increase Quantitative Proteome Coverage. *Analytical Chemistry, 87*, 6319–6327.

16. Gu, J., Lu, C., Guo, J., Chen, L., Chu, Y., Ji, Y., & Ge, D. (2013). Prognostic significance of the IASLC/ ATS/ ERS classification in Chinese patients-A single institution retrospective study of 292 lung adenocarci noma. *Journal of Surgical Oncology, 107* (5), 474–480.

17. Ha, S. Y., Choi, S. J., Cho, J. H., Choi, H. J., Lee, J., Jung, K., Irwin, D., Liu, X., Lira, M. E., Mao, M., Kim, H. K., Choi, Y. S., Shim, Y. M., Park, W. Y., Choi, Y. L., & Kim, J. (2015). Lung cancer in never-smoker Asian females is driven by oncogenic mutations, most often involving EGFR. *Oncotarget, 6* (7), 5465–5474.

18. Hirsh, V. (2015). Next-generation covalent irreversible Kinase inhibitors in NSCLC: Focus on Afatinib. *BioDrugs, 29* (3), 167–183.

19. Hood, B. L., Darfer, M. M., Guiel, T. G., Furusato, B., Lucas, D. A., Ringeisen, B. R., Sesterhenn, I. A., Conrads, T. P., Veenstra, T. D., & Krizman, D. B. (2005). Proteomic analysis of formalin-fixed prostate cancer tissue. *Molecular & Cellular Proteomics, 4*, 1741–1753.

20. Hu, H., Pan, Y., Li, Y., Wang, L., Wang, R., Zhang, Y., Li, H., Ye, T., Zhang, Y., Luo, X., Shao, L., Sun, Z., Cai, D., Xu, J., Lu, Q., Deng, Y., Shen, L., Ji, H., Sun, Y., & Chen, H. (2014). Oncogenic mutations are associated with histological subtypes but do not have an independent

prognostic value in lung adenocarcinoma. *Journal of OncoTargets and Therapy, 7*, 1423–1437.

21. Jänne, P. A., Yang, J. C., Kim, D. W., Planchard, D., Ohe, Y., Ramalingam, S. S., Ahn, M. J., Kim, S. W., Su, W. C., Horn, L., Haggstrom, D., Felip, E., Kim, J. H., Frewer, P., Cantarini, M., Brown, K. H., Dickinson, P. A., Ghiorghiu, S., & Ranson, M. (2015). AZD9291 in EGFR inhibitor-resistant non-small-cell lung cancer. *The New England Journal of Medicine, 372* (18), 1689–1699.
22. Jemal, A., Bray, F., Center, M. M., Ferlay, J., Ward, E., & Forman, D. (2011). Global cancer statistics. *CA: A Cancer Journal for Clinicians, 61*, 69–90.
23. Kato, Y., Nakamura, H., Tojo, H., Nomura, M., Nagao, T., Kawamura, T., Kodama, T., Ohira, T., Ikeda, N., Fehniger, T., Marko-Varga, G., Nishimura, T., & Kato, H. (2015). A proteomic profiling of lasermicrodissected lung adenocarcinoma cells of early lepidic-types. *Clinical and Translational Medicine, 4,* e24. doi: 10.1186/s40169-015-0064-3.
24. Kawamura, T., Nomura, M., Tojo, H., Fujii, K., Hamasaki, H., Mikami, S., Bando, Y., Kato, H., & Nishimura, T. (2010). Proteomic analysis of laser-microdissected paraffin-embedded tissues: (1) Stage-related protein candidates upon non-metastatic lung adenocarcinoma. *Journal of Proteomics, 73*, 1100–1110.
25. KEGG PATHWAY Database. (2016). Available from http://www.genome.jp/kegg/pathway.html. 16 Jan 2016.
26. Koike, T., Yamato, Y., Asamura, H., Tsuchiya, R., Sohara, Y., Eguchi, K., Mori, M., Nakanishi, Y., Goya, T., Koshiishi, Y., Miyaoka, E., & Japanese Joint Committee for Lung Cancer Registration. (2009). Improvements in surgical results for lung cancer from 1989 to 1999 in Japan. *Journal of Thoracic Oncology, 4,* 1364–1369.
27. LaBaer, J. (2012). Improving international research with clinical specimens: 5 achievable objectives. *Journal of Proteome Research, 11*, 5592–5601.
28. Leone, A. (2013). Highly sensitive detection of EGFR T790M mutation in pre-TKI specimens of EGFRmutated NSCLC: In cis, In trans, or a different clone? *Journal of Thoracic Oncology, 8* (3), e26–e27.
29. Malm, J., Végvári, A., Rezeli, M., Upton, P., Danmyr, P., Nilsson, R.,

Steinfelder, E., & Marko-Varga, G. (2012). Large scale biobanking of blood – The importance of high density sample processing procedures. *Journal of Proteomics, 76*, 116–124.

30. Mar, N., & Vredenburgh, J. J. (2015). Dual HER2 blockade in non-small cell lung cancer Harboring a HER2 mutation. *Connecticut Medicine, 79* (9), 531–535.
31. Marko-Varga, G. (2011). BioBanking – The Holy Grail of novel drug and diagnostic developments. *Journal of Clinical Bioinformatics, 13*, e14.
32. Marko-Varga, G., Végvári, A., Welinder, C., Rezeli, M., Edula, G., Svensson, K., Belting, M., Laurell, T., & Fehniger, T. E. (2011). Clinical protein science: utilization of biobank resources and examples of current applications. *Journal of Proteome Research, 11*, 5124–5134.
33. Michalski, A., Cox, J., & Mann, M. (2011). More than 100,000 detectable peptide species elute in single shotgun proteomics runs but the majority is inaccessible to data-dependent LC-MS/MS. *Journal of Proteome Research, 10*, 1785–1793.
34. Mitsudomi, T., Morita, S., Yatabe, Y., Negoro, S., Okamoto, I., Tsurutani, J., Seto, T., Satouchi, M., Tada, H., Hirashima, T., Asami, K., Katakami, N., Takada, M., Yoshioka, H., Shibata, K., Kudoh, S., Shimizu, E., Saito, H., Toyooka, S., Nakagawa, K., Fukuoka, M., & Group, W. J. O. (2010). Gefitinib versus cisplatin plus docetaxel in patients with non-smallcell lung cancer harbouring mutations of the epidermal growth factor receptor (WJTOG3405): An open label, randomised phase 3 trial. *The Lancet Oncology, 11* (2), 121–128.
35. Miyauchi, E., Inoue, A., Kobayashi, K., Maemondo, M., Sugawara, S., Oizumi, S., Isobe, H., Gemma, A., Saijo, Y., Yoshizawa, H., Hagiwara, K., Nukiwa, T., & North-East Japan Study Group. (2015). Efficacy of chemotherapy after first-line gefitinib therapy in EGFR mutation-positive advanced non-small cell lung cancer-data from a randomized Phase III study comparing gefitinib with carboplatin plus paclitaxel (NEJ002). *Japanese Journal of Clinical Oncology, 45* (7), 670–676.
36. Nakamura, H., & Saji, H. (2014). A worldwide trend of increasing primary adenocarcinoma of the lun. *Surgery Today, 44* (6), 1004–1012.
37. Nakamura, K., Nishio, K., Nishimura, T. (Eds.) & Kato, H (Ed. supervisor). (2012). Clinical Proteomics . Tokyo: Kanehara & Co. Nishimura T., Tojo H. (2014). Mass spectrometry-based protein

sequencing platforms. In G. Marko-Varga (Ed.), Genomics and proteomics for clinical discovery and development (pp. 69–99). (Wang, X (series Ed.) *Translational Bioinformatics 6). Dordrecht: Springer.*

38. Nomura, M., Fukuda, T., Fujii, K., Kawamura, T., Tojo, H., Kihara, M., Bando, Y., Gazdar, A. F., Tsuboi, M., Oshiro, H., Nagao, T., Ohira, T., Ikeda, N., Gotoh, N., Kato, H., Marko-Varga, G., & Nishimura, T. (2011). Preferential expression of potential markers for cancer stem cells in large cell neuroendocrine carcinoma of the lung. An FFPE proteomic study. *Journal of Clinical Bioinformatics, 1*, e23.
39. Nowell, P., & Hungerford, D. (1960). A minute chromosome in chronic granulocytic leukemia. *Science, 132*, 1497.
40. Oxnard, G. R., Nguyen, K. S., & Costa, D. B. (2014). Germline mutations in driver oncogenes and inherited lung cancer risk independent of smoking history. *Journal of the National Cancer Institute, 106* (1), djt361. doi: 10.1093/jnci/djt361.
41. Panchaud, A., Jung, S., Shaffer, S. A., Aitchison, J. D., & Goodlett, D. R. (2011). Faster, quantitative, and accurate precursor acquisition independent from ion count. *Analytical Chemistry, 83*, 2250–2257.
42. Physikron Mass Spectrometry Systems. (2013). Available from http://www.physikron.com/technology/. 15 Jan 2016.
43. Prieto, D. A., Hood, B. L., Darfl er, M. M., Guiel, T. G., Lucas, D. A., Conrads, T. P., et al. (2005). Liquid Tissue™: Proteomic profiling of formalin-fixed tissues. *BioTechniques, 38*, S32–S35.
44. RCSB- Research Collaboratory for Structural Bioinformatics. (2015). Protein Data Bank. Available from http://www.rcsb.org/pdb/home/home.do. 16 Jan 2016.
45. Reactome Pathway Database. (2016). Available from www.reactome.org/. 15 Jan 2016.
46. Sasaki, H., Shimizu, S., Tani, Y., Maekawa, M., Okuda, K., Yokota, K., Shitara, M., Hikosaka, Y., Moriyama, S., Yano, M., & Fujii, Y. (2012). RET expression and detection of KIF5B/RET gene rearrangements in Japanese lung cancer. *Cancer Medicine, 1* (1), 68–75.
47. Scagliotti, G. V., Longo, M., & Novello, S. (2009). Nonsmall cell lung cancer in never smokers. *Current Opinion in Oncology, 21*, 99–104.
48. Semenza, G. (2010). Defining the role of hypoxiainducible factor 1 in cancer biology and therapeutics. *Oncogene, 29,* 625–634.
49. Sequist, L. V., Rolfe, L., & Allen, A. R. (2015). Rociletinib in EGFR-

mutated non-small-cell lung cancer. The New England *Journal of Medicine, 373* (6), 578–579.

50. Shaw, A. T., Camidge, D. R., Engelman, J. A., Solomon, B. J., Kwak, E. L., Clark, J. W., Salgia, R., Shapiro, G., Bang, Y. J., Tan, W., Tye, L., Wilner, K. D., Stephenson, P., Varella-Garcia, M., Bergethon, K., Iafrate, A. J., Ou, S-H. I. (2012). Clinical activity of crizotinib in advanced non-small cell lung cancer (NSCLC) harboring ROS1 gene rearrangement, ASCO Annual Meeting. *Journal of Clinical Oncology, 30* (Suppl) abstract 7508.
51. Shtivelman, E., Hensing, T., Simon, G. R., Dennis, P. A., Otterson, G. A., Bueno, R., & Salgia, R. (2014). Molecular pathways and therapeutic targets in lung cancer. *Oncotarget, 5* (6), 1392–1433.
52. Siegel, R., Ward, E., Brawley, O., & Jemal, A. (2011). Cancer statistics, the impact of eliminating socioeconomic and racial disparities on premature cancer deaths. *CA: A Cancer Journal for Clinicians, 6*, 212–236.
53. Soda, M., Choi, Y. L., Enomoto, M., Takada, S., Yamashita, Y., Ishikawa, S., Fujiwara, S., Watanabe, H., Kurashina, K., Hatanaka, H., Bando, M., Ohno, S., Ishikawa, Y., Aburatani, H., Niki, T., Sohara, Y., Sugiyama, Y., & Mano, H. (2007). ldentification of the transforming EML4-ALK fusion gene in non-smallcell lung cancer. *Nature, 448* (7153), 561–566.
54. STRING-The Search Tool for the Retrieval of Interacting Genes/Proteins database. Available from http://stringdb. org/. 15 Jan 2016.
55. Sun, S., Schiller, J. H., & Gazdar, A. F. (2007). Lung cancer in neversmokers – A different disease. *Nature Reviews Cancer, 7* (10), 778–790.
56. The data base of Japanese Ministry of Health, Labor and Welfare. (2013) Available from http://www.mhlw.go.jp/toukei/saikin/hw/jinkou/geppo/nengai09/kekka3.html. 15 Jan 2016.
57. The data base of National Cancer Institute at the National Institute of Health. (2013). Available from http://www.cancer.gov/cancertopics/types/lung. 15 Jan 2016.
58. The National Lung Screening Trial Research Team. (2011). Reduced lung-cancer mortality with low-dose computed tomographic screening. *The New England Journal of Medicine, 365*, 395–409.
59. Thress, K. S., Paweletz, C. P., Felip, E., Cho, B. C., Stetson, D.,

Dougherty, B., Lai, Z., Markovets, A., Vivancos, A., Kuang, Y., Ercan, D., Matthews, S. E., Cantarini, M., Barrett, J. C., Jänne, P. A., & Oxnard, G. R. (2015). Acquired EGFR C797S mutation mediates resistance to AZD9291 in non-small cell lung cancer harboring EGFR T790M. *Nature Medicine, 21* (6), 560–562.

60. Toh, C. K., Wong, E. H., Lim, W. T., Leong, S. S., Fong, K. W., Wee, J., & Tan, E. H. (2004). The impact of smoking status on the behavior and survival outcome of patients with advanced non-small cell lung cancer: A retrospective analysis. *Chest, 126*, 1750–1756.
61. Tomlins, S. A., Laxman, B., Varambally, S., Cao, X., Yu, J., Helgeson, B. E., Cao, Q., Prensner, J. R., Rubin, M. A., Shah, R. B., Mehra, R., & Chinnaiyan, A. M. (2008). Role of the TMPRSS2-ERG gene fusion in prostate cancer. *Neoplasia, 10* (2), 177–188.
62. Travis, W. D. (2010). Advances in neuroendocrine lung tumors. *Annals of Oncology, 21* (Suppl. 7), vii 65–vii 71.
63. Travis, W. D., Brambilla, E., Müller-Hermelink, H. K., & Harris, C. C. (Eds.). (2004). Genetics of tumors of lung, pleura, thymus and heart (pp. 37–38). Lyon: IARC Press.
64. Travis, W. D., Brambilla, E., Noguchi, M., Nicholson, A. G., Geisinger, K. R., Yatabe, Y., et al. (2011). International association for the study of lung cancer/American thoracic society/european respiratory society international multidisciplinary classification of lung adenocarcinoma. *Journal of Thoracic Oncology, 6* (2), 244–285.
65. Travis, W. D., Colby, T. V., Corrin, B., Shimosato, Y., & Brambilla, E. (1999). Introduction. In *Histological Typing of Lung and Pleural Tumours* (pp. 1–19). Berlin/ Heidelberg: Springer.
66. Travis, W. D., Linnoila, R. I., Tsokos, M. G., Hitchcock, C. L., Cutler, G. B., Jr., Nieman, L., Chrousos, G., Pass, H., & Doppman, J. (1991). Neuroendocrine tumors of the lung with proposed criteria for large-cell neuroendocrine carcinoma. An ultrastructural, immu nohistochemical, and fl ow cytometric study of 35 cases. *The American Journal of Surgical Pathology, 15*, 529–553.
67. Varlotto, J. M., Medford-Davis, L. N., Recht, A., Flickinger, J. C., Schaefer, E., Zander, D. S., & DeCamp, M. M. (2011). Should large cell neuroendocrine lung carcinoma be classified and treated as a small cell lung cancer or with other large cell carcinomas? *Journal of Thoracic Oncology, 6* (6), 1050–1058.

68. Végvári, Á., Rezeli, M., Döme, B., Fehniger, T. E., & Marko-Varga, G. (2011a). Translation science for targeted personalized medicine treatments. In S. Sanders (Ed.), Selected presentations from the 2011 Sino- American symposium on clinical and translational medicine (pp. 36–37). Washington, DC: Science/ AAAS.

69. Végvári, Á., Welinder, C., Lindberg, H., Fehniger, T. E., & Marko-Varga, G. (2011b). Biobank resources for future patient care: Developments, principles and concepts. *Journal of Clinical Bioinformatics, 1*, e24.

70. Weinstein, I. B. (2001). Addiction to oncogenes—The Achilles heal of cancer. *Science, 297* (5578), 63–64.

71. Wisniewski, J. R., Ostasiewicz, P., & Mann, M. (2011). High recovery FASP applied to the proteomic analysis of micro dissected formalin fixed paraffin embedded cancer tissues retrieves known colon cancer markers. *Journal of Proteome Research, 10*, 3040–3049.

72. Wu, X, Fang, X, Zhu, Z, & Wang, X. (2014) Clinical Bioinformatics: A new emerging science of biomarker development. In Marko-Varga, G. (Ed.), Genomics and proteomics for clinical discovery and development (pp. 175–191), (Wang, X (series Ed.), *Translational Bioinformatics 6). Dordrecht: Springer.*

73. Yoshizawa, A., Motoi, N., Riely, G. J., Sima, C. S., Gerald, W. L., Kris, M. G., Park, B. J., Rusch, V. W., & Travis, W. D. (2011). Impact of proposed IASLC/ATS/ERS classification of lung adenocarcinoma: prognostic subgroups and implications for further revision of staging based on analysis of 514 stage I cases. *Modern Pathology, 24* (5), 653–664.

74. Yoshizawa, A., Sumiyoshi, S., Sonobe, M., Kobayashi, M., Fujimoto, M., Kawakami, F., suruyama, T., Travis, W. D., Date, H., & Haga, H. (2013). Validation of the IASLC/ATS/ERS lung adenocarcinoma classification for prognosis and association with EGFR and KRAS gene mutations: Analysis of 440 Japanese patients. *Journal of Thoracic Oncology, 8* (1), 52–61.

CHAPTER 8

PROTEOGENOMICS FOR ENVIRONMENTAL MICROBIOLOGY

CONTENTS

8.1. INTRODUCTION

Environmental microbiology is a cornerstone for basic knowledge concerning how life has looked on the earth, changed, and adapted to very varied and life-threatening conditions, in addition to being a favorable source of novel biotechnological products. Certainly, massive territories on earth are still generally unexplored regarding biodiversity, although they are surely colonized by life. At present, environmental microbiology is rapidly developing with novel instruments developed for perceiving life in situ and analyzing connections between organisms and their environments. Hyperspectral microscopes, atomic force microscopes, scanning transmission X-ray microscopes, and MS-based imagers are only some of the biophysical methods with promising prospective applications currently introduced to microbiology laboratories (Behrens et al., 2012). This field also paybacks from various methods and approaches for discovering the molecular mechanisms of isolated organisms at the closest level and, by extension, for evaluating the interactions between organisms in a given ecosystem (Christie-Oleza et al., 2012; Christie-Oleza et al., 2013). Definitely, next-generation sequencers, which permit sequencing nucleic acids at an extraordinary speed and depth, are by far the most significant sharers to the change in our awareness of microbial life. Today, the diversity of living cells can be evaluated rapidly and pretty reliably by amplifying 16S RNA-encoding genes and high-throughput sequencing. The genome of new isolated organisms can be also recognized at a comparatively low cost, and as a consequence, such projects can be now executed regularly in most microbiology laboratories.

Present research programs profit from the sequencing of various individual genomes or complex mixtures of environmental organisms to extensively enlarge the angle of comparison between samples (Yooseph et al., 2007). This weighty technology-driven tendency in environmental microbiology has led to a storm of genomic and metagenomic data (Armengaud et al., 2011).

The sum of data presently produced by next-generation sequencers is so great that it can be only analyzed with automated software programs or devoted pipelines. While the gathering of reads is frequently thought to lead to a confident genome sequence, owing to the high-sequence coverage acquired, the structural and functional annotation of the resultant sequences is by far the complex process, although it is typically taken for granted (Armengaud, 2009; Poptsova et al., 2010). In this chapter, we demonstrate data concerning annotation mistakes and list the hurdles in genome and proteome database curation.

We provide how proteogenomics, the utilization of large high-throughput proteomic datasets to improve genome annotation, may produce more reliable annotation methods. We also discuss probable drawbacks concerning accuracy and confidence. Furthermore, we demonstrate how proteogenomics-inspired approaches are growing and have led to new discoveries in the field of environmental microbiology.

8.2 AN EVALUATION OF ERRORS OCCURRING DURING GENOME SEQUENCING AND ANNOTATION

The current body of knowledge concerning genomes and proteomes is a collection of data collected numerous years ago with inaccurate techniques and high-throughput data collected more lately with enhanced methodologies that are fallible, however. A large count of nucleotide sequences presently stored in the DDBJ/EMBL/GenBank common database and accessible through the NCBI portal still comprise a comparatively high number of errors. Because only the scientists who primarily deposited the sequences and their annotations have the right to adjust these data, most are "as is." Their annotations, concerning gene structure and function, are inaccurate in nature, as many elements have yet to be exposed and characterized. This imprecise curated repository is a good example of the hurdles scientists have to take care of. Because genomes are now annotated by automated systems based on comparative genomics, a novel genome that does not possess any resemblances with already known genomes will be more liable to errors (Devos et al., 2001).

Certainly, it was demonstrated that the two first *Deinococcus* species sequences, *Deinococcus radiotolerans* BAA-816 (White et al., 1999) and *Deinococcus geothermalis* DSM1130 (Makarova et al., 2007), comprised various annotation errors when a third *Deinococcus* genome sequence, i.e., *Deinococcus deserti*, was established (de Groot et al., 2009. This is against the general concept where annotation errors are preserved and spread when the novel, comparatively close genomes are annotated (Armengaud et al., 2011).

Is there a reliable evaluation of errors in recent databases? Proteogenomics, i.e., the annotation of genomes with the aid of experimental proteomics data, which will be reviewed later here, allows for the documentation of the nature of errors in genome annotation and their global prevalence. In the case of *D. deserti*, a bulky proteomic analysis, where 1348 proteins were perceived from 11 129 unique tryptic peptides recorded employing MS/MS, was the basis for the genome annotation. Unexpectedly, 11 loci had been annotated in the mistaken reading orientation. Additionally, the products of 15 genes that were not expected by the annotation program were perceived. While only 40% of the theoretical proteome was revealed, the total count of missing genes is roughly 2% at the minimum.

Nevertheless, because small polypeptides are less possible to be perceived by MS and the annotation of the corresponding genes is difficult, the true percentage is perhaps much higher. More recently, the recognition of the *N*-termini of *D. deserti* proteins on a very large scale was executed employing a targeted method (Baudet et al., 2010). In this study, 664 *N*-terminal peptides were observed, allowing for the correction of 63 translation initiation codons in the *D. deserti* genome. Therefore, the ratio of incorrectly expected initiations is relatively appreciable (10%). Comparable results have been described for numerous models, such as *Salmonella typhimurium* (Anson et al., 2011), *Shigella flexneri* (Zhao et al., 2011), *Ruegeria pomeroyi* (Christie-Oleza et al., 2012), and *Pseudomonas fluorescens* (Kim et al., 2009).

8.3. PROTEOGENOMICS IS TARGETING AT BETTER ANNOTATION OF GENE STRUCTURES

8.3.1. Annotation needs Expert Validation

Detailed proteomics analysis of a particular sample depends on prefractionation of the complex protein pool, proteolysis by trypsin into peptides, and the utilization of high-resolution tandem mass spectrometers tied to high-pressure LC systems for resolving complex peptide mixtures (Armengaud, 2013). The resultant recognition of the proteins existent in the sample provides evidence of their synthesis. The easiest proteogenomics approach is founded on the clarification of recorded MS/MS spectra against a database in which an in-silico translation and digestion of genetic sequences are employed to provide all the probable polypeptide sequences encoded in the genome (Jaffe et al., 2004).

Most prokaryotic genomes are straightly translated succeeding the universal code, so their proteogenomics annotation is fairly direct.

Nevertheless, this excessively simplistic approach may not precisely mirror what happens in vivo. For instance, parasitic elements termed inteins may exist inserted into some protein sequences. Inteins are pieces of numerous hundred amino acids that are translated in-frame with the parasitized protein pieces and cut out themselves by an autocatalytic protein-splicing method, rejoining the residual fragments of the protein with a peptide bond result in the mature, functional protein. These elements have been revealed in all three domains of life and in viruses. Occasionally, the so-named "split intein" ligates the segments from two different genes, leading to a single mature protein. An example of this is the case for the catalytic subunit alpha of the DNA polymerase III DnaE from cyanobacteria (Caspi et al., 2003).

Luckily, these events are accounted uncommon and may be looked for precisely by nucleotide blast or comprehensive analysis of proteogenomics coverage. In eukaryotic genomes, the protein-coding data is broken up by the existence of numerous introns. Thus, proteogenomics strategies should integrate peptides in exon regions in addition to peptides precisely symbolic of exon junctions.

Another thing to put into consideration is the presence of frame shifts linked to recoding events (Antonov et al., 2013). While authentic indel mutations result in gene pseudogenization and therefore protein truncation and loss of function, revocable indel mutations can take place at specific sites in some members of the population, resulting in heterogeneity that subsequently helps the survival of those cells. Additionally, programmed frame shifts can take place either during transcription, i.e., programmed transcriptional realignments or translation, i.e., programmed ribosomal frame shifts. Again, these events are uncommon in living cells but common in viruses and transposons and can be followed using specific methods. The interrupted coding sequences revealed in a given genome sequence that has been previously annotated should be cautiously valued to define whether they signify authentic mutations or sequencing errors (Deshayes et al., 2007).

8.3.2. Proteogenomics Methods

To discover whether a specific ORF includes the data for a genuine polypeptide synthesized by living cells, the typical proteogenomics method for a prokaryotic genome is illustrated in Figure 8.1. It comprises three steps:

- six-frame translation of the nucleotide sequence to generate a STOP to STOP ORF database consisted of all ORF sequences opening after a STOP and finish at the next in-frame STOP,

- MS/MS spectra recording and task to the STOP to STOP ORF database,
- Authentication of perceived ORFs with peptides specific to these sequences and appraisal of the agreeing initiation codons (Figure 8.1). Choosing the possible STOP to STOP ORFs will considerably decrease the noise, as otherwise the ORF database will be overcrowded with unlikely polypeptide sequences. Exclusion of ORFs with a length below an assumed number of amino acids (typically 50) is logical. It is also likely to decrease the database by eliminating all the sequences that are not protein-like in structure (Blakeley et al., 2012). Nevertheless, this procedure may eliminate rare but remarkable genuine protein sequences from the database. Additionally, such an approach may be not productive due to the existence of an extended sequence at the *N*-terminus of most of the STOP to STOP ORFs that are artificially presented and does not agree to the mature protein sequence. A two-step database search has been lately planned to raise sensitivity in peptide recognition in proteogenomics or metaproteomics (Jagtap et al., 2013).

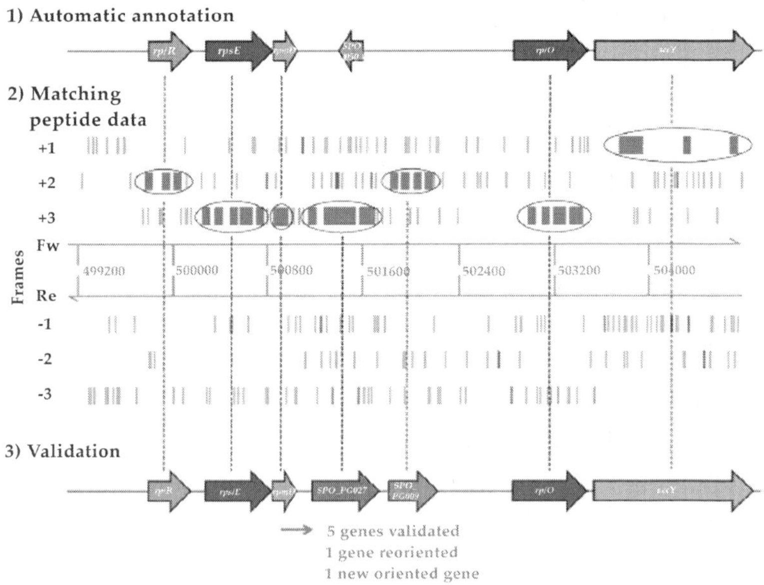

Figure 8.1. Proteogenomics mapping illustrated using the *Ruegeria pomeroyi* locus.

The most recent tools for assigning the MS/MS spectra take into consideration the probability of semi-specific protease cleavage at the *N*-terminus of the protein and correctly recognize this specific peptide caused by either the initiation of translation or the maturation of the peptide signal for export. While the first bioinformatic methods were custom scripts settled by some research groups for their own utilization, more incorporated tools have been presently published, such as the Proteogenomics Mapping Tool (Sanders et al., 2011), Genomic Peptide Finder (Specht et al., 2011), Minomics (Brouwer et al., 2009), and VESPA (Peterson et al., 2012) software workflows, which allow incorporation of proteomics and transcriptomics information with genome location organizes as well as BLAST analyses. Instead, a special approach is built on the understanding of each MS/MS spectra to extract the partial amino acid sequence of the peptide (Ferro et al., 2008). These peptide sequence tags are later mapped onto the nucleic acid sequences. Loci in which multiple contiguous tag hits are based are later analyzed to evaluate the coding sequence as an entire gene unit and outline the mature protein sequence. This method is well adjusted to eukaryotic sequences for which the existence of numerous introns renders the six-frame ORF translation inappropriate. A fully automated software program, named PepLine, has been used and evaluated for the annotation of the *Arabidopsis thaliana* envelope chloroplast proteome (Ferro et al., 2008).

8.4. *N*-TERMINOMICS IS THE TOOL TO SUCCESS

8.4.1. Unexpected Translation Starts

Ribosomes begin the process of protein synthesis from the initiation codon, but this codon only is not adequate to start the process. Initiation factors, and adjacent sequences are needed in this complexly controlled step of protein synthesis, but universal sequences are not clear inspite of the large body of literature on this subject (Myasnikov et al., 2009). The secondary structure of mRNA preferentially excludes entrée to most of the non-initiator methionine codons while leaving the true initiation site open.

The final identification of the initiation site by the ribosome only takes place with the cumulative and cooperative specificity of numerous ribosomal ligand-recognition sites (Nakamoto et al., 2009). AUG is recognized as the most frequent start codon and is read as methionine or, in bacteria, as formylmethionine. Two alternate initiation codons are frequently cited in most literature: GUG and UUG, which routinely code for the incorporation of valine and leucine residues, correspondingly, but as a start codon, they are translated into methionine or formylmethionine. In a large number of cases, the initiation codon will be AUG, GUG only being utilized in a comparatively low count of cases, and UUG being used unusually. Neverthless, this frequent utilization may be extremely different in some organisms. In a detailed multi-strategy proteomic study, Yamazaki et al. (2006) demonstrated that in the hyperthermophilic crenarchaeon *Aeropyrum pernix K1*, the translation of proteins in the greatest count of cases began from UUG (52%), with AUG (28%) and GUG (20%) far behind. Furthermore, other initiation codons have been revealed but are largely abandoned by genome annotators. Baudet et al. (2010) have shown by *N*-terminome-oriented proteogenomics that CUG, AUU, AUC, and AUA codons, which are supposed to encode for either leucine or isoleucine amino acids, can be employed as nonconventional translation begins within the same organism. Figure 8.2 shows a strong example of analysis by MS of a noncanonical AUC beginning codon for the *dnaA* translation in *D. deserti*.

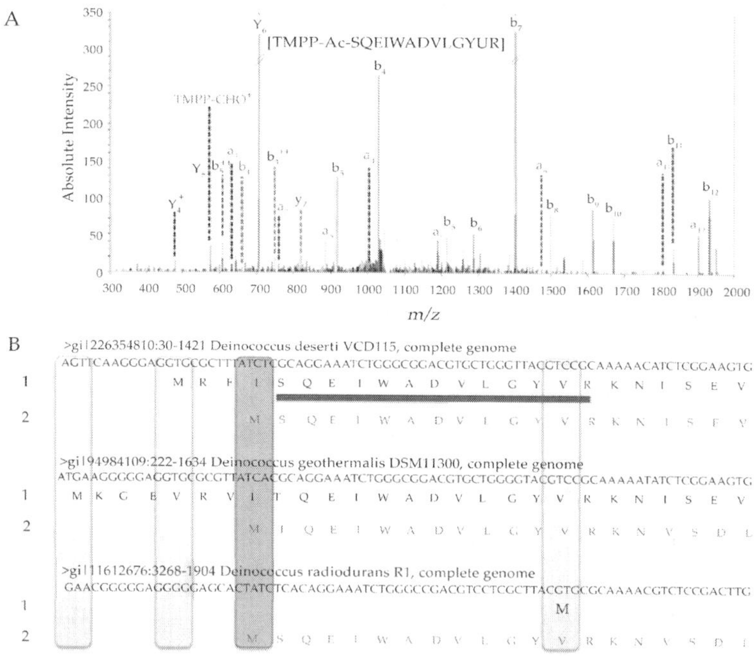

Figure 8.2. Discovery of a nonconventional translational start, AUC, for DnaA in Deinococcus.

8.4.2. Experimental Approaches for Systematic *N*-Terminomics

As a result of the alternate initiation beginning and numerous probable PTMs, automated annotation software programs show problems in correctly describing *N*-termini. In bacteria, the *N*-formylmethionine residue is post-translationally processed by a deformylase. In addition, the *N*-terminal methionine is removed from the protein by a methionine aminopeptidase when the second residue shows a small lateral chain, i.e., Gly, Ala, Ser, Cys, Thr, Pro, or Val (Hirel etal., 1989). Additional probable modification of the protein *N*-termini is *N*-acetylation by specific acetylases. It was presently reported through the compilation of data from almost 60 different proteomes that *N*-terminal acetylation takes place chiefly on Ala and Ser residues, bare after Met removal by an aminopeptidase (Bonissone et al., 2013). In addition, some proteins are exported and specific peptide signals have been more systematically demarcated (Payne et al., 2012). Their removal along translocation across the membrane for secretion is executed by signal peptidases, leading to protein products shorter than those encoded by the genome (Romine, 2011).

Compounding annotation difficulties, high-throughput shotgun proteomics does not widely cover the whole sequence of most proteins. This leads to low coverage of the *N*-terminal-most peptides. Some scientists have suggested methods to precisely enrich *N*-terminal peptides. These tools all depend on chemical modification of the *N*-termini before proteolysis with trypsin, which creates novel unmodified *N*-termini. The modified peptides are enriched with affinity chromatography or diagonal combined fractional diagonal chromatography. In this latter situation, the method includes four steps: acetylation to hinder free amino groups in the proteins, proteolysis with trypsin, reverse-phase chromatography to resolve the resultant peptides, which are reacted with a chemical that acts in response with the *N*-terminus of the internal peptides, and a second reverse-phase chromatography to distinct these novel modified peptides from the genuine*N*-terminal peptides (Staes et al, 2011).

This method has been currently applied to the *N*-terminome of the green sulfur bacterium *Chlorobaculum tepidum* (Kouyianou et al., 2012) and two *Coxiella burnetti* strains, the etiological agent of Q fever (Papadioti et al., 2012). Additional potent method to systematically list the *N*-termini of proteins is selective derivatization with *N*-Tris (2,4,6-trimethoxyphenyl) phosphonium acetyl (TMPP), a labeling reagent that augments the hydrophobic feature of the product, enhances its ionization, and adjusts the fragmentation feature by adding a positive charge. This tool has been applied for the wide identification of *N*-terminal peptides from *Mycobacterium smegmatis* (Gallien et al., 2009) and *D. deserti* (Baudet et al, 2010). It was suggested that the isotopic labeling of proteins and dimethylation of the *N*-termini with formaldehyde and sodium cyanoborohydride, tailed by digestion and an indirect purification with a high molecular weight dendritic polyglycerol aldehyde polymer which binds internal tryptic and C-terminal peptides which by now have free *N*-terminal alpha amines (Kleifeld et al., 2011). Nevertheless, activities to improve these "*N*-terminomics" tools are still required to get a wide-ranging view, the image being at the moment limited to no more than one-third of the theoretical *N*-terminome. Quantitative analysis of *N*-terminome is still quite a few; despite it may give a clearer view of the initiation of protein translation (Keller et al., 2010). In addition, new PTMs at the *N*-termini of proteins may occur, as presently evidenced in yeast where ten *N*-propionylated *N*-termini were noticed by MS (Foyn et al., 2013).

8.5. SHOULD WE TRUST IN PROTEOGENOMICS RESULTS?

To escape any novel sources of annotation mistakes and following criticisms, proteogenomics reannotations should be as reliable as to maximum. Special metrics should be developed in the coming future to evaluate the reliability of proteogenomics findings. Till now, research scientists executing proteogenomics have been careful, utilizing highly conservative features for re-annotation. Their novel expectations have been further affirmed with the assistance of sequence resemblances from strongly related genomes whenever achievable (De Souza et al., 2011; Kelkar et al., 2011; Zhao et al., 2011).

Nevertheless, the generalization of this method, the utilization of ultra-large datasets, or the extension to multiple organisms evaluated at once may result in the under-correction of annotation mistakes. At least four factors should be taken into account when checking confidence: the beginning nucleic acid sequence material, the experimental MS/MS data, the statistical development of both the MS/MS assignment and the proteogenomics information extraction, and finally, the proteogenomics data repository.

8.5.1. What Is the Nature of the Starting Nucleic Acid Sequence Material?

The quality of both the sequencing process and "the getting all together" is significant to escape any misleading interpretations of the proteomic information. In fact, numerous research groups have utilized proteogenomics to recognize the existence of sequencing mistakes that were additionally affirmed and adjusted by re-sequencing those specific loci (Deshayes et al., 2007; Christie-Oleza et al., 2012). While genomes sequenced several years ago may comprise a somehow high count of mistakes, this number is hugely declined today due to the high redundancy acquired with the modern next-generation DNA sequencers.

The count of errors for a small bacterial genome is predicted to be low, though dissimilarities may exist reliant on the genome GC content. Nevertheless, the situation may be quite different for large eukaryotic genomes. First, the scope of some fungal or algal genomes is so large that even a little mistake rate will result in a high number of errors overall. Additionally, the count and size of repetitive sequences in these genomes favor sequence mistakes and assembly errors.

In addition, populations of multicellular organisms are much more various than prokaryotic isolates. Whether the reference genome has been recognized for a precise member of this population or a combination of representatives is also significant. Moreover, an uncompleted genome sequence may be the raw beginning material, comprising in this situation a comparatively high count of contigs or scaffolds (Nanduri et al., 2005; Sun et al., 2006). Precise methods can be applied to evaluate the confidence of the sequencing stage with proteogenomics information. For instance, a search of the MS/MS spectral library utilizing the mistakes-tolerant mode accessible in the MASCOT software will emphasize non-silent nucleotide sequencing mistakes. Nevertheless, this method will not recognize mistakes hidden by the degeneracy of the genetic code.

8.5.2. How Confident We Should Be in the Experimental MS/MS Data and How Comprehensive is the Dataset?

Present tandem mass spectrometers are capable of producing large MS/MS datasets with less than 5 ppm, and frequently even sub-ppm, mass error on the parent ion (Wenger et al., 2010), therefore significantly decreasing the false-positive recognition of peptides. Nevertheless, the MS/MS spectra datasets presently utilized for most proteogenomics experiments are enormously large. For instance, a total of 10.9 million spectra have been gathered from the model legume *Medicago truncatula*; these spectrums translated to 78 647 distinct peptide sequences, which were evaluated as regarding proteogenomics (Volkening et al., 2012). Over 1.1 million MS/MS spectrums have been analyzed for the marine alpha proteobacterium *pomeroyi* grown under more than 30 different conditions to augment the covering of the theoretical proteome (Christie-Oleza et al., 2012).

While these datasets should be as complete as imaginable for a comprehensive analysis, the profusion of experimental data results in an appreciable global false-positive rate as regarding protein identification, even with very conservative factors for assigning spectrum to peptide sequences. Classically, 1 million MS/MS spectra allocated with a p value below 0.01 will result in 10,000 false-positive peptide recognition and therefore a high count of false detections. We should consider avoiding false detections, as much work may be done afterwards on the revision of proteins that never were! When extracting a precise item from the amassed data, a careful analysis of the whole false-positive rate as regarding protein recognition should be executed on the comprehensive pool of information. As a bigger database will yield a higher false positive rate, eukaryotic proteogenomics may need a better level of confidence in the allocation of MS/MS spectra. To decrease the density of the bioinformatics data handling, Helmy et al. (2012) currently suggested that the MS/MS spectra should be searched compared to the protein database and the spectra linked to recognized peptides should be eliminated to generate a smaller group of residual MS/MS spectra for searching compared to the nucleotide sequence database. Another solution is to group similar MS/MS spectra to allocate only the preeminent spectra (Flikka et al., 2007). Authenticating the peptide allocations with added physico-chemical information, such as molecular weight of the protein if known or elution time from chromatography, will lead to augmented confidence.

8.5.3. Should We Demand Specific Metrics for the Repository Of Proteogenomics Data?

Data handling and statistics for the MS/MS allocations in proteogenomics are acknowledged as present state-of-the art shotgun protocols next to consensus endorsements (Kinsinger et al., 2012). Regarding proteogenomics data extraction, data processing may fluctuate powerfully reliant on the authors and the scope of their studies. A new gene can be authenticated when at least two specific peptides designate the presence of its product. Most authors also investigate for resemblances between the resultant protein or nucleic acid sequences and closely linked proteins or genomes, correspondingly. Nevertheless, these data are either existing in supplementary information and are not always accessible after publication. Additionally, the comparative genomics-based approaches employed to authenticate findings are insufficiently illustrated in some cases. Because further biochemical authentication of the new detected protein is time overwhelming, such additional evidence have never been systematically offered. The well-illustrated, systematic authentication of revised annotations would be a key for development in the field of proteogenomics.

A new tendency in proteogenomics is the adding of transcriptomics data. With the expansion of high-throughput RNAseq, new pieces of experimental proof can be mapped onto the proteogenomics background (Denoeud et al., 2008). Nevertheless, many loci may be transcribed from diverse transcriptional initiation sites, producing biological blare buffering the total RNA pool, and only the translated and mature protein noticed by MS should be considered as regarding protein coding gene structures. In fact, an enormous count of long noncoding RNA entities has been identified, their functions being linked to the regulation of gene expression (Yoon et al., 2012).

It still appears a long shot to give the methods its good standards and discuss special metrics to evaluate the confidence of proteogenomics results. Nevertheless, it is a good approach to put all the experimental data in suitable repositories, permitting these data to be utilized for additional meta-analysis by other research teams. Considering the proteogenomics data themselves, special databases have been suggested, such as the rice proteogenomics database OryzaPG-DB (Helmy et al., 2012) and the CIPRO 2.5 integrated repository of large-scale omics data from *Ciona intestinalis* (Endo et al., 2011). They are harmonizing to present knowledge platforms for proteins, such as UniProtKB/Swiss-Prot (Boutet et al., 2007) or NeXtProt (Gaudet et al., 2013). Proteogenomics information should be transferred to these platforms to be accessible to the entire scientific society.

8.6. EXTENDING THE NOTIONS OF PROTEOGENOMICS

8.6.1. Earlier Eras

While the initial idea of "proteogenomics mapping" was set on its chief streaming, i.e., the annotation of genomes with the aid of proteomic information based on matched peptide and nucleic acid sequences (Jaffe et al., 2004), the proteogenomics era is moving in the way of the integration of more data, such as RNAseq and sequence resemblance information. Gupta et al. suggested the stimulating idea of comparative proteogenomics, putting into consideration evolutionary restrictions (Gupta et al., 2008). In this work, three *Shewanella* bacteria, i.e., *S. oneidensis*, *S. frigidimarina*, and *S. putrefaciens*, were cutivated under dissimilar situations and their proteomes were evaluated. Hundreds of translations beginning sites were established on the basis of the supposition that the matching codons should be exceedingly conserved among the huge majority of closely connected species. The use of dissimilar arrangements of MS data gathered from the three species was utilized to confidently recognize more proteins because the same protein seen with only one peptide in three diverse species is more statistically probable than three false positive dealings. Nevertheless, this method has not yet been accredited in other groups of bacteria, possibly because it is time overwhelming. We presently watched that closely connected strains may not have whole conserved traits as regarding *N*termini. Therefore, comparative proteogenomics findings should be investigated carefully.

8.6.2. New Proteogenomics-Flavored Approaches

The concept "proteogenomics" is a mix of proteomics and genomics, and therefore is also utilized by some research teams to illustrate studies where both proteomics and genomics are being used even if the link between the two fields is not firmly speaking linked to genome annotation. Another side for proteogenomics is the integrated used of genomic and proteomic data in projects targeting for answering a chief biological question. In these examples, the only drive is not overall genome annotation, but somewhat a pursuit that requires the inspection of specific annotation imprecisions or errors. For instance, an MS-based approach was presently suggested to investigate new bacterial isolates of a given genus. The method was developed with the genus *Ruegeria* as case model (Christie-Oleza et al., 2013). The methodology depends on the MALDI-TOF feature of the most plentiful small basic proteins from these isolates and the discovery of special biomarkers that are discoverable only if their sequences are precisely defined. Proteogenomics was utilized to state the character of the most reliable biomarkers, in addition to whether these biomarkers are post-translationally adjusted at their *N*-termini. Their most possible masses were then anticipated. This approach, where bottom-up and top-down methods were gathered, seemed quite vigorous and was used to recognize a new *Ruegeria* strain isolated from water sampled from the Mediterranean Sea. The recognition, executed by a simple acquisition of the whole-cell MALDI-TOF profile, was further affirmed by 16S RNA sequencing.

Composite proteomics was suggested some years ago as a tactic to execute proteomics on an unsequenced *Campylobacter* model, which is strictly linked to other sequenced organisms (Fagerquist, 2007). A special database that took benefit of the corresponding sequences was used. The searched proteins show sequences that share conserved segments with the homologous sequences existing in the database and have new rations that encompass mutations existing in one of the homologous sequences. As shown in Figure 8.3, specific combination of the diverse mutations found in the homologous sequences may yield the searched sequence. This proteogenomic-inspired method permitted the creation of sequences of a limited count of biomarkers (Fagerquist, et al., 2007). While this approach is beneficial, it has been limited to only a few proteins and has not yet been executed at the entire genome scale. Similarly, Castellana et al. (2010) suggested a process called template proteogenomics with the same target of beginning with an inaccurate database to allocate MS/MS spectra to peptide sequences.

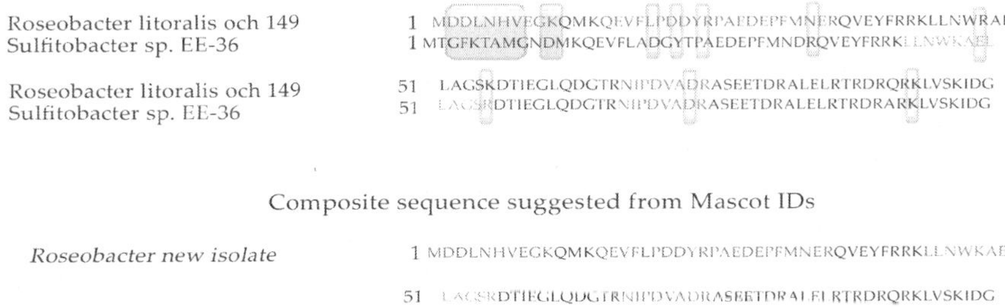

Figure 8.3. A strategy for composite proteomics.

The augmenting performance of mass spectrometers and the overflow of metagenomic data may be exploited in the near time to investigate metaproteomes, i.e., the entire proteome content of a given microcosm (Wilmes et al., 2009). Several research teams are presently linking forces to face these tasks, which are made enormously difficult due to the big variety of life on earth. Automation of the experimental and bioinformatics methods will permit the investigation of bigger collections of organisms (Venter et al., 2011). Such schemes should appear more commonly in the near time. Without hesitation, proteogenomics is an encouraging novel field where multi-omics data will be incorporated into a better considerate of life's close mechanisms.

8.7. POTENTIALS IN ENVIRONMENTAL MICROBIOLOGY

Proteogenomic investigations have already shared to numerous features of microbiology. Current studies have emphasized important proteins, PTMs, and pathways in extremophilic archaea and bacteria, naturally happening biofilms, engineered bioremediation applications, and microbial pathogens. These clarifications help to disentangle microbial procedures and relationships between associates of bacterial consortia and are directly pertinent to ongoing research and technological expansion. A proteogenomic study of an acid mine drainage biofilm community (Ram et al., 2005) revealed the position of a cytochrome, Cyt579, which was then isolated and identified (Singer et al., 2008), affirming the findings of the proteogenomic study. Additional investigation showed the presence of sequence alternatives and PTMs, comprising *N*-terminal cleavage, of this and other cytochromes over the sequence of biofilm development (singer et al., 2010). These proteins are vital for electron transport in metallo-respiration. Comparable proteogenomic examination of biofilms complicated in microbial fuel cells could explain electron transport phenomena used by anoderespiring bacteria, data that might aid in the design of more efficient fuel cells. Genes linked to radiotolerance and other genomic characteristics were exposed in proteogenomic studies of *Thermococcus gammatolerans* (Zivanovic et al., 2009) and *D. deserti* (Toueille et al., 2012).

These studies have shown unusual characteristics of these thrilling life forms, such as the use of noncanonical translation initiation codons. These interpretations challenge assumptions about basic cell mechanisms, resulting in unexpected findings. Proteogenomics was also utilized to observe metabolic and community shifts in *Geobacter* populations that may aid in managing uranium bioremediation (Wilkins et al., 2009). The method employed by Wilkins et al. (2009) leverages *Geobacter* isolate genome sequences and MS/MS to obviate the necessity for metagenomic sequencing of complex environmental samples. Changes in the observed proteins were linked to changes in the composition and metabolic activities of the community, which are further connected to the effects of biostimulation in the subsurface. Such research work can emphasize vital enzymes that could be used as microbial biomarkers to observe bioremediation steps, especially when an order of responses is necessary to fully transform a pollutant to a non- toxic metabolite, e.g., the reduction of trichloroethene to ethane (Ziv-El et al., 2012).

The detection and annotation of supplementary bioremediation-related genes would enable the search for new bioremediation agents and possibly explain the evolution of metabolic abilities previously known to be nonexistent. Different proteins were recognized as being important to fatty acid metabolism, the TCA cycle, and sugar metabolism in phosphorus-removing bacteria utilized in activated sludge treatment (Wilmes et al., 2009). Sludge digestion is a vital industrial process that relays on the act of microbial communities; presently, molecular techniques such as quantitative PCR and clone libraries are utilized to evaluate the character and function of digester populations (Lee et al., 2011). The existence of special species or genes, such as those accountable for phosphorus removal, can then be employed as a factor in models to expect the act of the system. Additional proteogenomic studies could recognize proteins whose quantitation will strengthen models for nutrient removal (Tang et al., 2013), the transformation of pharmaceuticals and personal care products (Heidler et al., 2009), and prediction of the fate of persistent organic pollutants (Heidler et al., 2009) in these workflows. The proteogenomic reannotation of a *Salmonella* species modified numerous start sites and recognized previously missed genes, comprising new genes potentially implicated in pathogenesis (Ansong et al., 2011). Likewise, reentering the genomes of three *Yersinia* species resulted in the reannotation of numerous genes, comprising virulence-associated factors (Schrimpe-Rutledge et al., 2012). The finding of previously overlooked vital or pathogenesis-related genes and proteins can supply novel targets for antibiotics, as well as novel biomarkers for the recognition of pathogenic organisms.

These proteogenomic studies have improved our basic comprehension of evolutionary procedures, our capability to accomplish microbial communities for engineering applications, and our ability to understand, and therefore combat, microbial pathogens. Given the prevalence and industrial significance of non culturable or obligate syntrophic microorganisms and the inherent errors of culture-based or targeted molecular methods, omics-scale studies are vital. However, mistakes in automated gene annotation are inevitable and non-negligible. The adding of systems level proteomics investigation to already largely implemented genomics techniques is therefore highly wanted if not absolutely vital.

REFERENCES

1. Ansong, C., Tolic, N., Purvine, S. O., Porwollik, S. et al. (2011). Experimental annotation of post-translational features and translated coding regions in the pathogen salmonella typhimurium. *BMC Genomics*, *12*, 433.
2. Antonov, I., & Baranov, P. (2013). Borodovsky,M., GeneTack database: genes with frameshifts in prokaryotic genomes and eukaryotic mRNA sequences. *Nucleic Acids Res.*, *41*, D152–D156.
3. Armengaud, J. (2009). A perfect genome annotation is within reach with the proteomics and genomics alliance. *Curr. Opin. Microbiol*, *12*, 292–300.
4. Armengaud, J. (2013). Microbiology and proteomics, getting the best of both worlds! *Environ. Microbiol*, *15*, 12–23.
5. Armengaud, J., Bland, C., Christie-Oleza, J. A., & Miotello, G. (2011). Microbial proteogenomics, gaining ground with the avalanche of genome sequences. *J. Bacteriol. Parasitol*, *S3*, 001.
6. Baudet, M., Ortet, P., Gaillard, J. C., Fernandez, B. et al. (2010). Proteomics-based refinement of *Deinococcus deserti* genome annotation reveals an unwonted use of noncanonical translation initiation codons. *Mol. Cell. Proteomics*, *9*, 415–426.
7. Behrens, S., Kappler, A., & Obst, M. (2012). Linking environmental processes to the in-situ functioning of microorganisms by high-resolution secondary ion mass spectrometry (NanoSIMS) and scanning transmission X-ray microscopy (STXM). *Environ. Microbiol*, *14*, 2851–2869. 2011, *10*, M111.011627.
8. Blakeley, P., Overton, I. M., & Hubbard, S. J. (2012). Addressing statistical biases in nucleotide-derived protein databases for proteogenomic search strategies. *J. Proteome Res.* 2012, *11*, 5221–5234.
9. Bonissone, S., Gupta, N., Romine, M., & Bradshaw, R. A. (2013). Pevzner, P. A., *N*-terminal protein processing: a comparative proteogenomic analysis. *Mol. Cell. Proteomics*, *12*, 14–28.
10. Boutet, E., Lieberherr, D., Tognolli, M., Schneider, M., Bairoch, A. (2007). UniProtKB/Swiss-Prot. *Methods Mol. Biol*, *406*, 89–112.
11. Brouwer, R. W., van Hijum, S. A., & Kuipers, O. P. (2009). MINOMICS: visualizing prokaryote transcriptomics and proteomics data in a genomic context. *Bioinformatics*, *25*, 139–140.

12. Caspi, J., Amitai, G., Belenkiy, O., & Pietrokovski, S. (2003). Distribution of split DnaE inteins in cyanobacteria. *Mol. Microbiol*, *50*, 1569–1577.
13. Castellana, N. E., Pham, V., Arnott, D., Lill, J. R., & Bafna, V. (2010). Template proteogenomics: sequencing whole proteins using an imperfect database. *Mol. Cell. Proteomics*, *9*, 1260–1270.
14. Christie-Oleza, J. A., Fernandez, B., Nogales, B., Bosch, R., & Armengaud, J. (2012). Proteomic insights into the lifestyle of an environmentally relevant marine bacterium. *ISME J*, *6*, 124–135.
15. Christie-Oleza, J. A., Miotello, G., & Armengaud, J. (2012). Highthroughput proteogenomics of *Ruegeria pomeroyi*: seeding a better genomic annotation for thewholemarine *Roseobacter clade*. *BMC Genomics*, *13*, 73.
16. Christie-Oleza, J. A., Pina-Villalonga, J. M., Guerin, P., Miotello, G. et al. (2013). Shotgun nanoLC-MS/MS proteogenomics to document MALDI-TOF biomarkers for screening new members of the Ruegeria genus. *Environ. Microbiol*, *15*, 133–147.
17. de Groot, A., Dulermo, R., Ortet, P., Blanchard, L. et al. (2009). Alliance of proteomics and genomics to unravel the specificities of Sahara bacterium *Deinococcus deserti*. *PLoS Genet*, *5*, e1000434.
18. de Souza, G. A., Arntzen, M. O., Fortuin, S., Schurch, A. C. et al. (2011). Proteogenomic analysis of polymorphisms and gene annotation divergences in prokaryotes using a clustered mass spectrometry-friendly database. *Mol. Cell. Proteomics*, *10*, M110.002527.
19. Denoeud, F., Aury, J. M., Da Silva, C., Noel, B. et al. (2008). Annotating genomes with massive-scale RNA sequencing. *Genome Biol*, *9*, R175.
20. Deshayes, C., Perrodou, E., Gallien, S., Euphrasie, D. et al. (2007). Interrupted coding sequences in *Mycobacterium smegmatis*: authentic mutations or sequencing errors? *Genome Biol*, *8*, R20.
21. Devos, D., Valencia, A. (2001). Intrinsic errors in genome annotation. *Trends Genet*, *17*, 429–431.
22. Endo, T., Ueno, K., Yonezawa, K., Mineta, K. et al. (2011). CIPRO 2.5: *Ciona intestinalis* protein database, a unique integrated repository of large-scale omics data, bioinformatic analyses and curated annotation, with user rating and reviewing functionality. *Nucleic Acids Res*, *39*, D807–D814.

23. Fagerquist, C. K. (2007). Amino acid sequence determination of protein biomarkers of *Campylobacter upsaliensis* and *C. helveticus* by "composite" sequence proteomic analysis. *J. Proteome Res*, *6*, 2539–2549.

24. Fagerquist, C. K., Yee, E., & Miller, W. G. (2007). Composite sequence proteomic analysis of protein biomarkers of *Campylobacter coli, C. lari* and C. *concisus* for bacterial identification. *Analyst*, *132*, 1010–1023.

25. Ferro, M., Tardif, M., Reguer, E., Cahuzac, R. et al. (2008). PepLine: a software pipeline for high-throughput direct mapping of tandem mass spectrometry data on genomic sequences. *J. Proteome Res*, *7*, 1873–1883.

26. Flikka, K., Meukens, J., Helsens, K., Vandekerckhove, J. et al. (2007). Implementation and application of a versatile clustering tool for tandem mass spectrometry data. *Proteomics*, *7*, 3245–3258.

27. Foyn, H., Van Damme, P., Stove, S. I., Glomnes, N. et al. (2013). Protein *N*-terminal acetyltransferases act as *N*-terminal propionyltransferases in vitro and in vivo. *Mol. Cell. Proteomics*, *12*, 42–54.

28. Gallien, S., Perrodou, E., Carapito, C., Deshayes, C. et al. (2009). Ortho-proteogenomics: multiple proteomes investigation through orthology and a new MS-based protocol. *Genome Res*, *19*, 128–135.

29. Gaudet, P., Argoud-Puy, G., Cusin, I., Duek, P. et al. (2013). neXtProt: organizing protein knowledge in the context of human proteome projects. *J. Proteome Res*, *12*, 293–298.

30. Gupta, N., Benhamida, J., Bhargava, V., Goodman, D. et al. (2008). Comparative proteogenomics: combining mass spectrometry and comparative genomics to analyze multiple genomes. *Genome Res*, *18*, 1133–1142.

31. Heidler, J., & Halden, R. U. (2009). Fate of organohalogens in US wastewater treatment plants and estimated chemical releases to soils nationwide from biosolids recycling. *J. Environ. Monitor*, *11*, 2207–2215.

32. Helmy, M., Sugiyama, N., Tomita, M., & Ishihama, Y. (2012). Mass spectrum sequential subtraction speeds up searching large peptide MS/MS spectra datasets against large nucleotide databases for proteogenomics. *Genes Cells*, *17*, 633–644.

33. Helmy, M., Sugiyama, N., Tomita, M., & Ishihama, Y. (2012). The rice proteogenomics database OryzaPG-DB: development, expansion, and new features. *Front. Plant Sci*, *3*, 65.

34. Hirel, P. H., Schmitter, M. J., Dessen, P., Fayat, G., & Blanquet, S. (1989). Extent of *N*-terminal methionine excision from *Escherichia coli* proteins is governed by the side-chain length of the penultimate amino acid. *Proc. Natl. Acad. Sci. USA*, *86*, 8247–8251.

35. Jaffe, J. D., Berg, H. C., & Church, G. M. (2004). Proteogenomic mapping as a complementary method to perform genome annotation. *Proteomics*, *4*, 59–77.

36. Jagtap, P., Goslinga, J., Kooren, J. A., McGowan, T. et al. (2013). A two-step database search method improves sensitivity in peptide sequence matches for metaproteomics and proteogenomics studies. *Proteomics*, *13*, 1352–1357.

37. Kelkar, D. S., Kumar, D., Kumar, P., Balakrishnan, L. et al., Proteogenomic analysis of *Mycobacterium tuberculosis* by high resolution mass spectrometry. *Mol Cell Proteomics. 2011 Dec;10(12):M111.011627.*

38. Keller, U., Prudova, A., Gioia, M., Butler, G. S., & Overall, C. M. (2010). A statistics-based platform for quantitative *N*-terminome analysis and identification of protease cleavage products. *Mol. Cell. Proteomics*, *9*, 912–927.

39. Kim, W., Silby, M. W., Purvine, S. O., Nicoll, J. S. et al. (2009). Proteomic detection of non-annotated protein-coding genes in *Pseudomonas fluorescens* Pf0–1. *PLoS ONE*, *4*, e8455.

40. Kinsinger, C. R., Apffel, J., Baker, M., Bian, X. et al. (2012). Recommendations for mass spectrometry data quality metrics for open access data (corollary to the Amsterdam principles). *Proteomics*, *12*, 11–20.

41. Kleifeld, O., Doucet, A., Prudova, A., auf dem Keller, U. et al. (2011). Identifying and quantifying proteolytic events and the natural N terminome by terminal amine isotopic labeling of substrates. *Nat. Protoc.*, *6*, 1578–1611.

42. Kouyianou, K., De Bock, P. J., Colaert, N., Nikolaki, A. et al. (2012). Proteome profiling of the green sulfur bacterium *Chlorobaculum tepidum* by *N*-terminal proteomics. *Proteomics*, *12*, 63–67.

43. Lee, I. S., Parameswaran, P., & Rittmann, B. E. (2011). Effects of solids

retention time on methanogenesis in anaerobic digestion of thickened mixed sludge. *Bioresource Technol*, *102*, 10266–10272.

44. Makarova, K. S., Omelchenko, M. V., Gaidamakova, E. K., Matrosova, V. Y. et al. (2007). *Deinococcus geothermalis*: the pool of extreme radiation resistance genes shrinks. *PLoS ONE*, *2*, e955.
45. Myasnikov, A. G., Simonetti, A., Marzi, S., & Klaholz, B. P. (2009). Structure-function insights into prokaryotic and eukaryotic translation initiation. *Curr. Opin. Struct. Biol*, *19*, 300–309.
46. Nakamoto, T. (2009). Evolution and the universality of the mechanism of initiation of protein synthesis. *Gene*, *432*, 1–6.
47. Nanduri, B., Lawrence, M. L., Vanguri, S., Pechan, T., & Burgess, S. C. (2005). Proteomic analysis using an unfinished bacterial genome: the effects of subminimum inhibitory concentrations of antibiotics on *Mannheimia haemolytica* virulence factor expression. *Proteomics*, *5*, 4852–4863.
48. Papadioti, A., De Bock, P. J., Vranakis, I., Tselentis, Y. et al. (2012). Study of the whole cell lysate of two Coxiella burnetii strains using *N*-terminomics. *J. Proteome Res*, *11*, 3150–3159.
49. Payne, S. H., Bonissone, S., Wu, S., Brown, R. N. et al. (2012). Unexpected diversity of signal peptides in prokaryotes. *MBio*, *3*, e00339–12.
50. Peterson, E. S., McCue, L. A., Schrimpe-Rutledge, A. C., Jensen, J. L. et al. (2012). VESPA: software to facilitate genomic annotation of prokaryotic organisms through integration of proteomic and transcriptomic data. *BMC Genomics*, *13*, 131.
51. Poptsova, M. S., & Gogarten, J. P. (2010). Using comparative genome analysis to identify problems in annotated microbial genomes. *Microbiology*, *156*, 1909–1917.
52. Ram, R. J., Verberkmoes, N. C., Thelen, M. P., Tyson, G. W. et al. (2005). Community proteomics of a natural microbial biofilm. *Science*, *308*, 1915–1920.
53. Romine, M. F. (2011). Genome-wide protein localization prediction strategies for gram negative bacteria. *BMC Genomics*, *12*(Suppl 1), S1.
54. Sanders, W. S., Wang, N., Bridges, S. M., Malone, B. M. et al. (2011). The proteogenomic mapping tool. *BMC Bioinformatics*, *12*, 115.
55. Schrimpe-Rutledge,A. C., Jones, M. B., Chauhan, S., Purvine, S. O. et al. (2012). Comparative omics-driven genome annotation refinement:

application across Yersiniae. *PLoS ONE*, *7*, e33903.

56. Singer, S. W., Chan, C. S., Zemla, A., VerBerkmoes, N. C et al. (2008). Characterization of cytochrome 579, an unusual cytochrome isolated from an iron-oxidizing microbial community. *Appl. Environ. Microbiol.*, *74*, 4454–4462.
57. Singer, S.W., Erickson, B. K., VerBerkmoes, N. C., Hwang, M. et al. (2010). Posttranslational modification and sequence variation of redox-active proteins correlate with biofilm life cycle in natural microbial communities. *ISME J*, *4*, 1398–1409.
58. Specht, M., Stanke, M., Terashima, M., Naumann-Busch, B. et al. (2011). Concerted action of the new genomic peptide finder and AUGUSTUS allows for automated proteogenomic annotation of the *Chlamydomonas reinhardtii* genome. *Proteomics*, *11*, 1814–1823.
59. Staes, A., Impens, F., Van Damme, P., Ruttens, B. et al. (2011). Selecting protein *N*-terminal peptides by combined fractional diagonal chromatography. *Nat. Protoc*, *6*, 1130–1141.
60. Sun, J., Wang, W., Hundertmark, C., Zeng, A. P. et al. (2006). A protein database constructed from low-coverage genomic sequence of *Bacillus megaterium* and its use for accelerated proteomic analysis. *J. Biotechnol.*, *124*, 486–495.
61. Tang, Y., Ontiveros-Valencia, A., Feng, L., Zhou, C. et al. (2013). A biofilmmodel to understand the onset of sulfate reduction in denitrifying membrane biofilm reactors. *Biotechnol. Bioeng.*, *110*, 763–772.
62. Toueille, M., Mirabela, B., Guerin, P., Bouthier de la Tour, C. et al. (2012). A comparative proteomic approach to better define Deinococcus nucleoid specificities. *J. Proteomics*, *75*, 2588–2600.
63. Venter, E., Smith, R. D., & Payne, S. H. (2011). Proteogenomic analysis of bacteria and archaea: a 46-organism case study. *PLoS ONE*, *6*, e27587.
64. Volkening, J. D., Bailey, D. J., Rose, C. M., Grimsrud, P. A. et al. (2012). A proteogenomic survey of the *Medicago truncatula* genome. *Mol. Cell. Proteomics*, *11*, 933–944.
65. Wenger, C. D., McAlister, G. C., Xia, Q., & Coon, J. J. (2010). Sub-part-per-million precursor and product mass accuracy for high-throughput proteomics on an electron transfer dissociation-enabled orbitrap mass spectrometer. *Mol. Cell. Proteomics*, *9*, 754–763.

66. White, O., Eisen, J. A., Heidelberg, J. F., Hickey, E. K. et al. (1999). Genome sequence of the radioresistant bacterium *Deinococcus radiodurans* R1. *Science, 286*, 1571–1577.

67. Wilkins, M. J., Verberkmoes, N. C., Williams, K. H., Callister,S. J. et al. (2009). Proteogenomic monitoring of Geobacter physiology during stimulated uranium bioremediation. *Appl. Environ. Microbiol., 75*, 6591–6599.

68. Wilmes, P., & Bond, P. L. (2009) Microbial community proteomics: elucidating the catalysts and metabolic mechanisms that drive the Earth's biogeochemical cycles. *Curr. Opin. Microbiol., 12*, 310–317.

69. Yamazaki, S., Yamazaki, J., Nishijima, K., Otsuka, R. et al. (2006). Proteome analysis of an aerobic hyperthermophilic crenarchaeon, *Aeropyrum pernix* K1. *Mol. Cell. Proteomics., 5*, 811–823.

70. Yoon, J. H., Abdelmohsen, K., & Gorospe, M. (2012). Posttranscriptional gene regulation by long noncoding RNA. *J. Mol. Biol.*, doi: 10.1016/j.jmb.2012.11.024.

71. Yooseph, S., Sutton, G., Rusch, D. B., Halpern, A. L. et al. (2007). The Sorcerer II Global Ocean Sampling expedition: expanding the universe of protein families. *PLoS Biol., 5*, e16.

72. Zhao, L., Liu, L., Leng, W., Wei, C., Jin, Q. (2011). A proteogenomic analysis of *Shigella flexneri* using 2D LC-MALDI TOF/TOF. *BMC Genomics, 12*, 528.

73. Zivanovic, Y., Armengaud, J., Lagorce, A., Leplat, C. et al. (2009). Genome analysis and genome-wide proteomics of *Thermococcus gammatolerans*, the most radioresistant organism known amongst the Archaea. *Genome Biol.* 2009, *10*, R70.

74. Ziv-El, M., Popat, S. C., Parameswaran, P., Kang, D. W. et al. (2012). Using electron balances and molecular techniques to assess trichoroethene-induced shifts to a dechlorinating microbial community. *Biotechnol. Bioeng., 109*, 2230–2239.

CHAPTER 9

PROTEOGENOMICS FOR THE ANALYSIS OF HUMAN ANTIBODY REPERTOIRES

CONTENTS

9.1. INTRODUCTION

Since the breakthrough detection of the antibody (also denoted to as immunoglobulins, Ig) in blood serum, it has been well-described that a composite spectrum of different antibodies is present in the blood. This extensive spectrum is created by B-cell clones through a somatic recombination manner (Christiansen et al., 2007). Lately, numerous studies have recognized and determined the comparative abundance of the monoclonal antibodies (mAbs) contained within the serum pool that is provoked in reaction to vaccination or natural infection. To comprehend the humoral reactions, it is significant to attain information about the arrangement of the antigen-specific serum antibody repertoires, the characteristics (e.g., recognized epitopes, affinity) of the corresponding Ig, the connection between circulating Ig, and the existence of clonally expanded peripheral B-cells.

In the most recent years, the time of current genomics and proteomics is giving amazing new methods for investigating antibody repertoires. Next Generation Sequencing (NGS) permits the recognition of millions of B-cell receptor (BCR) sequences in a sole experiment. In addition, NGS techniques allow the investigation of the human antibody catalogue, not only to help in the detection of distinct antibodies possibly valuable as therapeutics, but also to broadly catalog the antibody sequences that are provoked during the adaptive immune reaction. In addition, enhancements made on NGS strategy have permitted the gaining of the endogenous variable light and heavy (VL and VH) pairs inside NGS datasets.

Therefore, owing to the sequencing of this paired VH: VL, the BCR catalog investigation has been successfully enhanced (Figure 9.1). Lastly, the combination of NGS methods and high resolution protein mass spectrometry (MS) has augmented the identification of serum antibodies (Lavinder et al., 2015). Though deep sequencing technologies for DNA and RNA have been progressed in the last decade, this genomic and transcriptomic data does not precisely mirror the real cellular status, as every sequence that is transcribed may not be translated (the translated percent of the genome is remarkably decreased in comparison to the transcribed portion) (Haider and Pal, 2013). Therefore, proteogenomics seems as the combinatorial method that utilizes proteomic data to increase and supplement the transcriptomic significance (Woo et al., 2014).

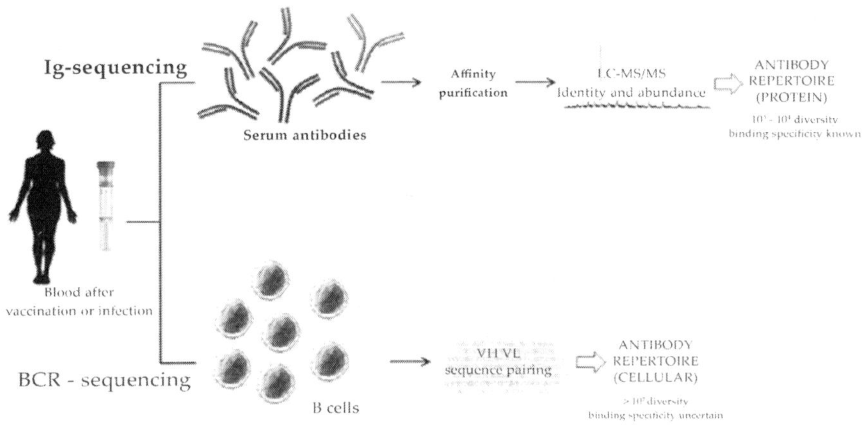

Figure 9.1. Sequencing of immunoglobulins (Ig) and B-cell receptor (BCR).

9.2. PROTEOGENOMICS

The chief target of this novel discipline, proteogenomics, is to identify and classify new peptides, combining the two strategies, proteomics and genomics. Therefore, genomic and transcriptomics data is utilized to create customized protein sequence databases that are employed as reference databases for MS data. Subsequently, proteomics can provide protein-level indication of gene expression (Castellana and Bafna, 2010).

As a result of the progress of new sequencing methods (*e.g.*, RNA-Seq, NGS) and the enhancements completed in the proteomics field, proteogenomics has shown a tremendous increase in consideration. To comprehend the ground of proteogenomics, it is important to primarily define the implicated categories in the field (*i.e.*, genomics, transcriptomics and proteomics). Proteomics is the wide-scale comprehensive investigation of proteins and search for the description of their structures and functions, among other characteristics. The concept was initiated in 1994 by Marc Wilkins as an equivalent to the notion of genomics. There are diverse proteomic approaches to acquire proteomic data but usually 'shotgun proteomics' is the designated method in which liquid chromatography (LC) and tandem mass spectrometry (MS/ MS) are gathered. To characterize peptides with these methods, it is essential to utilize a reference database of speculative protein sequences and it is in this idea where a novel plan is needed as numerous peptides do not exist in any of the reference databases. Numerous substitutes have lately appeared, comprising sequence tag-based database searching or *de novo* sequencing (Nesvizhskii, 2014). Nevertheless, these methods are incompetent for wide-scale investigations as their working times are pretty long and some consumables are quite expensive (Fullwood et al., 2009; Hert et al., 2008). By these methods, proteogenomics appears to be a sufficient substitute to recognize novel peptides.

Nesvizhskii has proposed some guidelines for proteogenomics investigations (Nesvizhskii, 2014):

- Make the customized protein sequence databases accessible to be utilized to recognize new peptides
- Search the peptides as compared to all chief reference databases and map to shared sample contaminants
- Illustrate the FDR estimation method
- Mark peptides mapping to various genome locations for peptide recognition by proteogenomics methods, it is significant that the gained MS/MS spectra are matched in comparison to a customized protein sequence database. There are various ways to advance such databases that we detail below.

(1) **Six-frame translation.** The sequence of nucleotides (DNA or RNA) can be divided into successive and non-consecutive triplets and read in six dissimilar reading frames reliant on the reading direction (5′ → 3′ or 3′ → 5′) and the beginning reading point in the triplet (1st, 2nd, or 3rd nucleotide). As long as the sequences of concern are translated in all 6 frames, the agreeing peptide/protein sequences for each frame are created (Winnenburg et al., 2008). There are numerous bioinformatic and computational approaches obtainable to automatically create these translations (*e.g.*, attain orf from the EMBOSS (European Molecular Biology Open Software Suite), ExPASy, Bioline). This method provides a massive variety of probable peptides and the tremendously large size of the resultant database represents a limitation. Additionally, most of the produced sequences do not actually happen, needing the formation of rules (*e.g.*, homology to known coding sequences, minimum length) for the assortment of the possible frames (Nesvizhskii, 2014).

(2) ***Ab initio* methods.** These methods permit the prediction of genes – harboring no resemblance to those formerly described – with a high specificity and sensitivity. There is an extensive range of informatics methods for *ab initio* gene prediction,

such as MetaGene, EasyGene, Glimmer, GeneMark, Augustus, or GeneID (Mathé et al., 2002). All these methodologies systematically search for *signals* and/or *content* denoting the existence of protein-coding sequences in the DNA. *Signals* are precise sequences that designate the intimate existence of a gene (*e.g.*, Pribnow box), while *content* is stated to the statistics characteristics fitting to the coding sequence (*e.g.*, statistics for stop codons) (Zhu et al., 2010).

(3) **Expressed sequence tag (EST) data.** Short sequences (~300 nucleotides) received from a cDNA sequence are known as expressed sequence tag (EST) and they signify parts of expressed genes. In public cataolgs, there is accessible a large count of ESTs produced by sequencing the 5′- or 3′- end of randomly sequestered cDNA clones. There were 74,186,692 ESTs public inputs registered in the EST database in the US National Center for Biotechnology Information (NCBI,January11, 2016) (http://www.ncbi.nlm.nih. gov/genbank/dbest/dbest_summary/) (The et al., 2011). ESTs produce peptide sequence examples in a more direct way when paralleled to *ab initio* gene prediction method, but a problem appears when utilizing the six-frame translation of this information due to the escalating augmentation in the size of the produced database (Nesvizhskii, 2014).

(4) **Annotated RNA transcripts.** As a substitute for producing protein sequences employing the six-frame translation from DNA, there is a substitute utilizing the three-frame translation of annotated RNA transcripts.

– **RNA-Seq data.** RNA-Seq can be thought as an enhanced model of ESTs in which the time and cost are considerably decreased when put in comparison to Sanger sequencing. In addition, its dynamic assortment is great due to the fact that sequencing reads mapping to distinctive regions is limitless. Nevertheless, the chief troublesome issue in this technique is the annotation of non-abundant genes as they are characterized by scarce reads (Roberts et al., 2011). This gene annotation method have permitted the illustration of many transcriptomes, from specific organisms, such as *Deinococcus deserti* (de Groot et al., 2014), to organs implicated in disorders (*e.g.*, the transcriptome of the brain endothelium from patients suffering from a cerebrovascular malfunction in ischemic stroke (Zhang et al., 2015).

As soon as the protein sequence database has been customized, it is essential to create the proteomics data. The operator can then recognize the peptides by gathering the two sources of data. One acknowledged approach to raise the count of recognized peptides is grounded on the application of various searching methods with the same dataset, gathered with rescoring of peptide recognitions. Additionally, it is important to evaluate the recognition confidence to eliminate the possibility of wrong recognitions at protein, peptide and spectrum levels. With this target, decoy sequences are involved in the search database at each investigation level. With this data, a false discovery rate (FDR) can be estimated for the calculation of global error rates. Classically, it is proposed that FDR at peptide-spectrum matches (PSM), peptide, and protein levels should be only around 1% which is interpreted as that 99% of the recognition does not match with decoy sequences from the search database and are reflected as correct recognitions. When considering the new peptides, it is essential to put in mind than they may show stronger proof than well-recognized peptides to be supposed as new peptides. Moreover, FDR estimation should be done distinctly for new and known peptides (Reiter et al., 2009; Elias and Gygi, 2010; Nesvizhskii, 2014).

The incorporation of –omics approach should yield new information about the performance of the cell and the link between the transcribed genes and their translated complements. Inspire of the promising findings, there are still various encounters that should be presented and dealt with to raise the quality of the biological data. Numerous studies have presently addressed this issue, showing promising outcomes as a result of the gathering different 'omics' methods. Below there is a discussion of some of these findings. The incorporation of proteomics and transcriptomics datasets was utilized to investigate a lymphoma B-cell line (Ramos) by Díez and colleagues (Díez et al., 2015). Proteomics information acquired utilizing a nanoUPLC-LTQ-Orbitrap Velos was gathered with the transcriptomic outlining of the Ramos cells demonstrating a 94% mixture in the proteins recognized by both 'omics' methods. A further investigation revealed 30% coverage of all protein- coding genes existing in the human genome.

Original datasets were handled to categorize the recognized proteins in various groups relaying on handling and confidence levels. In this manner, the created datasets were as follows:

- *intersection,* proteins methodically recognized in replicated experiments with 2 or more proteotypic peptides at protein FDR < 0.01

- *union,* proteins recognized in any experiment with 2 or more proteotypic peptides at protein FDR < 0.01 (iii) *maximum,* proteins detected in any experiment with 1 or more proteotypic peptides at protein FDR < 0.01 This categorization permits the choosing of proteins in function of the strictness level and exhibits the complementarity between methods to get a whole map of the investigated lymphoma B-cells.

McRedmond and colleagues (McRedmond et al., 2004) executed the incorporation of proteomics and genomics for the identification and description of platelets. Precisely, they recognized 82 secreted proteins and relating them to the transcriptome data acquiring 69% correlation between both 'omics' data. In addition, they suspected the existence of new proteins in the platelets. Aditional example is the research study executed by Günther et al. in which they demonstrated that the integration of 'omics' approaches could be utilized for dimension lessening and identification of examples of biomarkers in acute kidney transplant rejection (Günther et al., 2014).

9.3. ANALYSIS METHODS FOR SEQUENCING IMMUNOGLOBULIN GENES

The immunoglobulins (Ig), also named as antibodies, are Y-shaped proteins, chiefly generated by plasma cells (Figure 9.2). Their function is to detect and neutralize pathogenic antigens. Igs are comprising four chains: two indistinguishable light chains (L) and two indistinguishable heavy chains (H). In mammals, light chains can be categorized into two types: lambda and kappa. These chains are linked by disulfide bonds at dissimilar points of the structure. In addition, two parts can be distinct in antibodies: the variable region (V) and the constant region (C). The variable region contains the epitope binding site where the antigen is identified. Lastly, the structure of the Ig can be separated into the Fab region, for the antigen specificity, and the Fc region, for defining the class effect of the antibody (Schroeder and Cavacini, 2013).

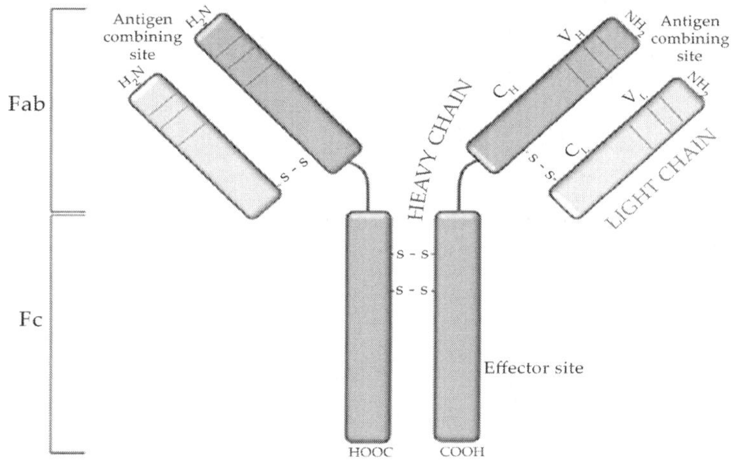

Figure 9.2. A representative structure of an immunoglobulin.

The enormous amount of dissimilar Ig proteins generated to counter the huge collection of antigens is produced owing to a multi-layer mechanism established by the immune system. The molecular sequence diversity of B and T cell receptors (BCR and TCR, correspondingly) defines their capability to bind to a huge variety of antigens. This diversity is produced by three distinct but harmonizing mechanisms: VDJ recombination, somatic hypermutation, and antigen selection (Ralph and Matsen, 2015).

The VDJ recombination process, found by Susumu Tonegawa (Hozumi and Tonegawa, 1976), includes the random selection of V, D, and J genes which will be additionally gathered in a procedure that omits some nucleotides from their sequences. The final sequence produced identifies the specificity against the antigen (these sequences are named as complementary determining regions, CDR) (Ralph and Matsen, 2015). Alternatively, somatic hypermutation is a procedure in which single base substitutions are produced, with infrequent deletions and insertions, in the variable region of the Ig, produced by the B-cells. The rate of somatic hypermutation in humans is 10 −5 –10 −3 mutations per base pair per production level. This procedure produces better specific antibodies (Li et al., 2004; Kostareli et al., 2012).

In addition, the antigen selection theory – developed by Burnet in the early 1940s – revealed that those B lymphocytes producing a specific antibody which is capable of blocking a specific antigen are activated to produce clones for antibody production (Jordan and Baxter, 2008). With all these procedures, the total diversity of produced Ig molecules is theoretically limitless. Then, their investigation should be concentrated on the thorough analysis of their sequences. With this aim, the Ig BLAST analysis method was progessed at NCBI. Its algorithm is grounded on the BLAST search algorithm comprising BCR-specific aspects:

- Reporting of the gene matches from the germline V, D, and J domains to the query sequence
- Annotation of the Ig domain
- Providing the V(D)J junction details
- Data on the rearrangement (inframe and out-of-frame) Special requirements are needed for identifying Ig sequences, since they are quite long (290 bases for V genes). In this manner, Ig BLAST has the Capability to advance various queries (<1000 sequences per batch) (Ye et al., 2009).

In addition, there is an online annotation approach named IMGT, which is referred as the global reference in immunoinformatics and immunogenetics. Its content contains information about Ig, TCR, and major histocompatibility complex (MHC), among others (Lefranc et al., 2015). Other candidates of germline databases comprise; IHMMune-align, VBASE2, and JoinSolv er (Ye et al., 2009).

9.4. DE NOVO PROTEIN SEQUENCING OF MONOCLONAL ANTIBODIES

Defining peptide amino acid sequences is probable in tandem MS. This procedure is known as "*de novo* peptide sequencing" and it has its starts in the Edman degradation procedure. Opposing to database searching, *de novo* sequencing permits the identification of novel peptides as it allocates fragment ions from a mass spectrum (Hughes et al., 2010). Nevertheless, the application of *de novo* sequencing for recognizing the sequences of antibodies is still an encounter. In these situations, MS/MS is not appropriate and it makes it essential to search for novel approaches. Traditional Edman degradation could be a choice, but it is a low-throughput and laborious strategy. However, Bandeira and colleagues have established a novel method known as Comparative Shotgun Protein Sequencing (CSPS) to recognize unknown proteins utilizing known proteins as templates in less than 72 h.

The operative of CSPS is grounded in a three stage sequence: alignment, assembly, and consensus procedure. At the alignment stage, spectral alignments from overlapping peptides are recognized. The gathering of spectral alignments into spectral contigs takes place at the assembly step, and these spectral contigs turn into protein contigs at the consensus step (Bandeira et al., 2008).

As illustrated in the Syd Labs web page (http://www.sydlabs.com/de-novo-antibody- sequencing-service-p58.htm), the overall tool for *de novo* antibody sequencing includes:

- Initial tryptic digestion tailed by LC-MS/MS
- Scale-up digestion with other enzymes (Asp-N, Glu-C) tailed by fragmentation optimization.
- Separation of antibody heavy chain and light chain if needed.
- Data analysis. A huge count of studies has implemented *de novo* sequencing to identify antibody sequences. For exmaple, Pham and collaborators illustrated the sequence of a full-length monoclonal antibody increased as a result against OX40 ligand, comprising heavy and light chain sequences. In this resrach study, they gathered Edman degradation and mass spectrometric analysis (Pham et al., 2006).

In turn, Resemann and colleagues identified the primary structure of a 13.6 kDa single heavy chain camelid antibody (V H H) utilizing top down mass spectrometric investigative analysis approach revealed that around 14 kDa proteins can be sequenced as a whole by MS (Resemann et al., 2010). Lastly, the gathering of high-resolution MS, *de novo* sequencing, and reverse engineering and chimerization tactics have permitted the purification and sequencing of antibodies consequent for ascites, precisely, the heavy and light chain sequences of the LT-3 F12 antibody (Castellana et al., 2011).

9.5. NATIVE MASS SPECTROMETRY

Native mass spectrometry is a newly introduced technology for the analysis of the native-like quaternary assemblies. Despite this method does not present molecular or atomic structure data, it is categorized by its various advantages, comprising unlimited and dynamic mass range, high sensitivity, speed selectivity, and accuracy. In addition, it permits the isolation of a specific sample within a mixed protein complex and the necessity of minor amounts of sample (minimum of 10 picomoles) (van Duijn, 2010). To summerize, this method can be reflected as an intermediate platform between interactomics and structural biology, though other technologies are required to fill the gaps (*e.g.*, nuclear magnetic resonance (NMR) spectroscopy, X-ray crystallography, and electron microscopy).

Native MS is also valuable for filtering structural models (Heck, 2008; van Duijn, 2010). For native MS, electrospray ionization (ESI) is one of the supreme utilized choices to formulate the sample for the additional MS analysis. To conserve the quaternary protein structure, an aqueous ammonium acetate solution has been added, since it is well-matched with the MS process. As regards the mass analyzer, it is essential that it has a high precision for ion identification (Heck, 2008; van Duijn, 2010). Native MS has permitted the study of proteasomes, RNA polymerase II and III, and protein complexes, among others. Regarding proteasomes, numerous researchers have utilized native MS to describe them in different species (*e.g.,, Methanosarcina thermophile, Thermoplasma acidophilum, Rhodococcus erythropolis*) giving information about their stoichiometries and masses (Heck, 2008). Comparable descriptions were done for RNA polymerases by Lorenzen and colleagues (Lorenzen et al., 2007) for polymerases from yeast.

Other studies have been concentrated on the investigation of monoclonal antibodies, comprising the assessment of antibody-antigen binding, structural features, dynamics, and interaction strengths. Rosati and colleagues have described monoclonal antibodies and one of the earlier encounters that they established was linked to the existence of N-linked glycosylation sites in each heavy chain as they are highly reliant of the sort of cells. By utilizing native MS in orbitrap analyzers, they were capable of performing the studies in a sensitive and fast manner (Rosati et al., 2012).

9.6. CLINICAL POTENTIALS OF PROTEOGENOMICS APPROACHES

Considering all these progresses in ProteoGenomics, the following step will focus on their implementation in the clinical field, for both diagnostics and prognostics. For example, enhancing the data about the structural particularities of Igs may aid as a forecast factor of the probable development of pathologies, as was demonstrated by Lomakin and colleagues (Lomakin et al., 2014) for multiple sclerosis disease. In this study, they searched for the "molecular signature" of viruses in the Ig catalog, discovering specific variable heavy and light chains from Ig against the myelin basic protein

(MBP). In addition, they demonstrated the cross-reactivity of this protein with the latent membrane protein 1 (LMP1) of Epstein-Barr virus which proposes that the accountable part of the antibody specificity is the light chain of the Ig. As regarding cancer research, proteogenomics emerges as a promising tool to better comprehend how cancer progression takes place. Therefore, the illustration of molecular alterations in cancer has been executed by deep genome sequencing (led by the International Cancer Genome Consortium and The Cancer Genome Atlas, TCGA). Nevertheless, the necessity of incorporating genotypes and phenotypes, together with the development of proteomics, has enhanced the founding of an international consortium (Clinical Proteomics Tumor Analysis Consortium, CPTAC) targeting the understanding the molecular basis of cancer (Faulkner et al., 2015). For example, the proteogenomics illustration of human colon and rectal cancers has presently been characterized, which detect prospective candidate biomarkers and therapeutic targets. These results were made probable owing to the linkage of global alteration at mRNA and protein levels (Zhang et al., 2014). Sequentially, Fanayan et al. incorporated proteomics (shotgun tool) and transcriptomics (RNA-Seq) methods to investigate human colon cancer cell lines (LIM1215, LIM1899, and LIM2405) within the chromosomecentric human proteome project (C-HPP). Their findings comprised potential markers for colorectal cancer, including nucleophosmin, ezrin, mortalin, and exportin, among others (Fanayan et al., 2013).

9.7. CONCLUSIONS

In order to ilustrate the wide count of Igs generated by the immune system, it is needed that new tools are developed to attain this target. Conventionally, genome-scale techniques have been implemented to raise the knowledge about Igs (*e.g.*, sequence of the antigen binding site, domains for epitope recognition). In this manner, various strategies have been successfully applied in the genomics area. Special consideration should be assumed to next generation sequencing (NGS) that permits illustrating millions of B-cell receptor sequences in a single experiment. However, current developments in the proteomics area have emphasized the requirement of novel perspectives when analyzing antibody catalogs. Therefore, gathering techniques – not only traditional genomics and transcriptomics techniques but also novel 'omics' areas, such as proteomics and metabolomics – could actually indicate a development in the clinical area where immunoglobulins and their properties are critical for the expansion of disease.

REFERENCES

1. Bandeira, N., et al. (2008). Automated de novo protein sequencing of monoclonal antibodies. *Nature Biotechnology, 26* (12), 1336–1338.
2. Castellana, N. E., et al. (2011). Resurrection of a clinical antibody: Template proteogenomic de novo proteomic sequencing and reverse engineering of an anti- lymphotoxin alpha antibody. *Proteomics, 11* (3), 395–405.
3. Castellana, N., & Bafna, V. (2010). Proteogenomics to discover the full coding content of genomes: A computational perspective. *Journal of Proteomics, 73* (11), 2124–2135.
4. Christiansen, M., et al. (2007). Chapter 4: Maritime transportation. In *Handbooks in Operations Research and Management Science,* 14(C) (pp. 189–284). Amsterdam: Elsevier.
5. de Groot, A., et al. (2014). RNA sequencing and proteogenomics reveal the importance of leaderless mRNAs in the radiation-tolerant bacterium Deinococcus deserti. *Genome Biology and Evolution, 6* (4), 932–948.
6. Díez, P., et al. (2015). Integration of Proteomics and Transcriptomics data sets for the analysis of B-cell lymphoma cell line in the context of the Chromosome- Centric Human Proteome Project. *J. Proteome Res. 14* (9):3530–40.
7. Elias, J. E., & Gygi, S. P. (2010). Target-decoy search strategy for mass spectrometry-based proteomics. *Methods in Molecular Biology (Clifton, NJ), 604* (2), 55–71.
8. Fanayan, S., et al. (2013). Proteogenomic analysis of human colon carcinoma cell lines LIM1215, LIM1899, and LIM2405. *Journal of Proteome Research, 12,* 1732–1742.
9. Faulkner, S., Dun, M. D., & Hondermarck, H. (2015). Proteogenomics: Emergence and promise. *Cellular and Molecular Life Sciences, 72* (5), 953–957. Available at: http://link.springer.com/10.1007/ s00018-015-1837-y.
10. Fullwood, M. J., et al. (2009). Next-generation DNA sequencing of paired-end tags (PET) for transcriptome and genome analyses. *Genome Research, 19* (4), 521–532.
11. Günther, O. P., et al. (2014). Novel multivariate methods for integration of genomics and proteomics data: Applications in a kidney transplant rejection study. *Omics: A Journal of Integrative Biology, 18* (11), 682–

695.

12. Haider, S., & Pal, R. (2013). Integrated analysis of transcriptomics and proteomic data. *Current Genomics, 14* (2), 91–110.
13. Heck, A. J. R. (2008). Native mass spectrometry: A bridge between interactomics and structural biology. *Nature Methods, 5* (11), 927–933.
14. Hert, D. G., Fredlake, C. P., & Barron, A. E. (2008). Advantages and limitations of next-generation sequencing technologies: A comparison of electrophoresis and non-electrophoresis methods. *Electrophoresis, 29,* 4618–4626.
15. Hozumi, N., & Tonegawa, S. (1976). Evidence for somatic rearrangement of immunoglobulin genes coding for variable and constant regions. *Proceedings of the National Academy of Sciences of the United States of America, 73* (10), 3628–3632.
16. Hughes, C., Ma, B., & Lajoie, G. A. (2010). De novo sequencing methods in proteomics. *Methods in Molecular Biology (Clifton, NJ), 604,* 105–121.
17. Jordan, M. A., & Baxter, A. G. (2008). Quantitative and qualitative approaches to GOD: The first 10 years of the clonal selection theory. *Immunology and Cell Biology, 86* (1), 72–9.
18. Kostareli, E., et al. (2012). Immunoglobulin gene repertoire in chronic lymphocytic leukemia: Insight into antigen selection and microenvironmental interactions. *Mediterranean Journal of Hematology and Infectious Diseases, 4* (1), e2012052.
19. Lavinder, J. J., et al. (2015). Next-generation sequencing and protein mass spectrometry for the comprehensive analysis of human cellular and serum antibody repertoires. *Current Opinion in Chemical Biology, 24,* 112–120.
20. Lefranc, M.-P., et al. (2015). IMGT®, the international ImMunoGeneTics information system® 25 years on. *Nucleic Acids Research, 43* (Database issue), D413–D422.
21. Li, Z., et al. (2004). The generation of antibody diversity through somatic hypermutation and class switch recombination. *Genes and Development, 18* (1), 1–11.
22. Lomakin, Y. a., et al. (2014). Heavy-light chain interrelations of MS-associated immunoglobulins probed by deep sequencing and rational variation. *Molecular Immunology, 62* (2), 305–314.
23. Lorenzen, K., et al. (2007). Structural biology of RNA polymerase

III: Mass spectrometry elucidates subcomplex architecture. *Structure (London, England: 1993), 15* (10), 1237–1245.

24. Mathé, C., et al. (2002). Current methods of gene prediction, their strengths and weaknesses. *Nucleic Acids Research, 30* (19), 4103–4117.
25. McRedmond, J. P., et al. (2004). Integration of proteomics and genomics in platelets: A profile of platelet proteins and platelet-specific genes. *Molecular & Cellular Proteomics, 3* (2), 133–144.
26. Nesvizhskii, A. I. (2014). Proteogenomics: Concepts, applications and computational strategies. *Nature Methods, 11* (11), 1114–1125. Available at: http:// www.pubmedcentral.nih.gov/articlerender.fcgi?artid=4392723&tool=pmcentrez&rendertype=abstract.
27. Pham, V., et al. (2006). De novo proteomic sequencing of a monoclonal antibody raised against OX40 ligand. *Analytical Biochemistry, 352* (1), 77–86.
28. Ralph, D., & Matsen, F. A. (2015). Consistency of VDJ rearrangement and substitution parameters enables accurate B cell receptor sequence annotation, *arXiv* (qbio.PE),1–40.
29. Reiter, L., et al. (2009). Protein identification false discovery rates for very large proteomics data sets generated by tandem mass spectrometry. *Molecular & Cellular Proteomics, 8* (11), 2405–2417.
30. Resemann, A., et al. (2010). Top-down de novo protein sequencing of a 13.6 kDa camelid single heavy chain antibody by matrix-assisted laser desorption ionization- time-of-fl ight/time-of-fl ight mass spectrometry. *Analytical Chemistry, 82* (8), 3283–3292.
31. Roberts, A., et al. (2011). Identification of novel transcripts in annotated genomes using RNA-Seq. *Bioinformatics, 27* (17), 2325–2329.
32. Rosati, S., et al. (2012). Exploring an orbitrap analyzer for the characterization of intact antibodies by native mass spectrometry. *Angewandte Chemie – International Edition, 51* (52), 12992–12996.
33. Schroeder, H. W., & Cavacini, L. (2013). Structure and function of immunoglobulins. *Journal Allergy Clinical Immunology, 125* (125), S41–S52.
34. Teh, S.-L., et al. (2011). Development of expressed sequence tag resources for Vanda Mimi Palmer and data mining for EST-SSR. *Molecular Biology Reports, 38* (6), 3903–3909.
35. van Duijn, E. (2010). Current limitations in native mass spectrometry

based structural biology. *Journal of the American Society for Mass Spectrometry, 21* (6), 971–978.

36. Winnenburg, R., et al. (2008). PHI-base update: Additions to the pathogen – Host interaction database. *Database, 36* (October 2007), 572–576.
37. Woo, S., et al. (2014). Proteogenomic database construction driven from large scale RNA-Seq data. *Journal of Proteome Research, 13* (1), 21–28.
38. Ye, X., Blonder, J., & Veenstra, T. D. (2009). Targeted proteomics for validation of biomarkers in clinical samples. *Briefings in Functional Genomics & Proteomics, 8* (2), 126–135.
39. Zhang, B., et al. (2014). Proteogenomic characterization of human colon and rectal cancer. *Nature, 513* (7518), 382–387.
40. Zhang, J., et al. (2015). Altered long non-coding RNA transcriptomic profiles in brain microvascular endothelium after cerebral ischemia. *Experimental Neurology, 277,* 162–170.
41. Zhu, W., Lomsadze, A., & Borodovsky, M. (2010). Ab initio gene identify cation in metagenomic sequences. *Nucleic Acids Research, 38* (12), 1–15.

CHAPTER 10

CONCLUDING MARKS

Early proteogenomics methods always faced the technical challenges driven by an overall low sensitivity of proteomics technology. The last years, however, have yielded great improvements in MS instrumentation, comprising new instrument types, alternative fragmentation mechanisms, and novel data acquisition strategies. These progressions, coupled with enhancement in protein separation and enrichment methods for proteomics applications, have permitted a significant augmentation in the depth of protein detection, approaching that of global transcriptome profiling studies. Substantial progressions have also been witnessed in the area of top-down proteomics – a technology that provides additional information highly beneficial for proteogenomics characterization.

A big challenge in proteogenomics has been always the shortage of a sufficient amount of proteomic data in the public domain essential to make a significant share to genome annotation efforts, in one side due to the "data hoarding" mentality common in the proteomics community in the early days. Nowadays, there is a clear inclination toward more open data sharing in proteomics, further strengthened by new needs from funding agencies. Consequently, an increasing count of proteomic datasets is currently reachable in public repositories.

Although the magnitude of RNA-Seq data is still higher than that of proteomic data, transcriptome data comprises elements not anticipated to include mature proteins (e.g., a large number of nonfunctional transcripts). Ribosome profiling data gives an evidence of translational activity and therefore can be utilized to detect new transcripts that are more probable to be protein-coding. Nevertheless, these data do not yield the clear evidence of expression of a stable, functional protein. Accordingly, in spite of the remarkable success of RNA-Seq and related technologies in unveiling the previously unidentified diversity of the genome, proteomic data exerts a key role in recognizing functional transcripts among the new transcripts nominated by genomics and transcriptomics technologies. Joint analyses utilizing multi-omics data should be substantially informative when done in close integration between the genomics and proteomics group, with biological experiments carefully planned to create paired genomics/proteomics datasets.

Looking at proteogenomics in a general context, queries endure as to what fraction of new alternative splice forms is translated into stable functional proteins vs. those that are susceptible to nonsense mediated decay or protein degradation instantly after translation. Further analysis can help recognize the variances between confirmed (at the protein level) and unconfirmed splice forms according to their secondary structure and sequence features. Current studies also suppose that SAVs could affect protein stability, probably illustrating the lower than expected rate of recognition of such variants in proteomic data. In addition, somatic variants have been established to have declined protein abundance in comparison to germline variants. These and other current studies comprising quantitative analysis of transcripts and protein expression data and integration with DNA variation yield appreciated visions into how the proteome is regulated using genetic effects. In summary, there is strong evidence that the field of proteogenomics will stay as a vivid area of research for the near future.

INDEX

G

H

I

L

M

N

O

P

R

S

T

W

15